WORKSHEETS
FOR CLASSROOM OR LAB PRACTICE

with contributions from

CHRISTINE VERITY

BASIC COLLEGE MATHEMATICS
ELEVENTH EDITION

Marvin Bittinger

Indiana University – Purdue University IN

Addison-Wesley
is an imprint of

Copyright © 2010, 2007, 2003 Pearson Education, Inc.
Publishing as Pearson Addison-Wesley, 75 Arlington Street, Boston, MA 02116.

ISBN-13: 978-0-321-62728-5
ISBN-10: 0-321-62728-8

3 4 5 6 BRR 12 11 10

Addison-Wesley
is an imprint of

www.pearsonhighered.com

Table of Contents

Name: Date:
Instructor: Section:

Chapter 1 WHOLE NUMBERS

1.1 Standard Notation

Learning Objectives
a   Give the meaning of digits in standard notation.
b   Convert from standard notation to expanded notation.
c   Convert between standard notation and word names.

Key Terms
Use the vocabulary terms listed below to complete each statement in Exercises 1–2. Terms may be used more than once.

   whole numbers   natural numbers

1.   The set 1, 2, 3, 4, 5,... is called the set of _____.

2.   The _____ include all the _____ as well as 0.

Objective a   Give the meaning of digits in standard notation.

What does the digit 6 mean in each number?

3.   478,621   3. _____

4.   362,908   4. _____

5.   4,906,357   5. _____

6.   6,382,015   6. _____

In the number 4,729,508, what digit names the number of:

7.   hundred thousands?   7. _____
8.   tens?   8. _____

Objective b Convert from standard notation to expanded notation.

Write expanded notation.

9. 9012

10. 46,238

11. 851,724

12. 3,085,926

13. 208,640

9. _____

10. _____

11. _____

12. _____

13. _____

Objective c Convert between standard notation and word names.

Write a word name.

14. 47

15. 3905

16. 24,096

17. 8,750,231

14. _____

15. _____

16. _____

17. _____

Write standard notation.

18. Three hundred seventy-six thousand, eight hundred forty

19. Two thousand, six hundred five

20. Thirteen billion

18. _____

19. _____

20. _____

2

Chapter 1 WHOLE NUMBERS

1.2 Addition

Learning objectives
a   Add whole numbers.
b   Use addition in finding perimeter.

Key Terms
Use the vocabulary terms listed below to complete each statement in Exercises 1–2.

sum   perimeter

1.   The _____ is the distance around an object.

2.   The _____ is the result of an addition.

Objective a   Add whole numbers.

Add.

3.   5 2 1
      +   4 6

3. _____

4.   4 9 2 3
        + 3 3 5 2

4. _____

5. $7226 + 540$

5. _____

6. $842 + 65,304$

6. _____

7.   7 6 5 4
        + 3 8 9 6

7. _____

8.   49,375
          +58,426

8. _____

9.   643
            926
            807
           +294

9. _____

Objective b   Use addition in finding perimeter.

Find the perimeter of the figure.

10.

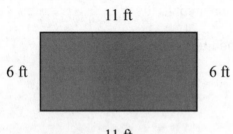

11 ft

6 ft   6 ft

11 ft

10. _____

11.

11. _____

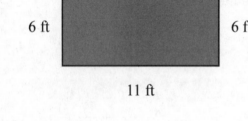

5 yd

3 yd

4 yd

12.

12. _____

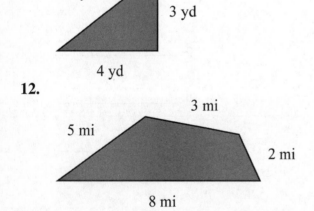

3 mi

5 mi

2 mi

8 mi

4

Chapter 1 WHOLE NUMBERS

1.3 Subtraction

Learning objectives
a   Subtract whole numbers.

Key Terms

Use the vocabulary terms listed below to complete each statement in Exercises 1–2.

difference   **subtrahend**

1. The _____ is the result of a subtraction.

2. In a subtraction sentence, the number being subtracted is the _____.

Objective a   Subtract whole numbers.

Subtract.

3.   9 3
        − 4 1

3. _____

4.   4 8 7 6
     − 2 3 5 4

4. _____

5.   5 4 3 8
     − 3 6 6 2

5. _____

6. $49,075 - 38,246$

6. _____

7. 430
  −168

7. _____

8. 705
  −398

8. _____

9. 23,000
  −14,928

9. _____

10. 728 − 643

10. _____

11. 5729
  −2865

11. _____

12. 29,042 − 1985

12. _____

13. 4433
  −2888

13. _____

14. 65,000
  −59,871

14. _____

6

Chapter 1 WHOLE NUMBERS

1.4 Multiplication

Learning objectives
a   Multiply whole numbers.
b   Use multiplication in finding area.

Key Terms
Use the vocabulary terms listed below to complete each statement in Exercises 1–2.

factor   product

1.   A _____ is the result of a multiplication.

2.   A number that we multiply is called a _____.

Objective a   Multiply whole numbers.

Multiply.

3.   2 6   3. _____
      × 3
      ———

4.   6 0 0   4. _____
      × 4 0
      ———

5.   6·507   5. _____

6.   8(2946)   6. _____

7.   4 3 6
       × _5 2_

7. _____

8.   8 5 9 3
       × _7 0 6 3_

8. _____

Objective b   Use multiplication in finding area.

What is the area of the region?

9.

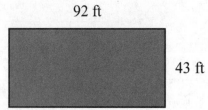

12 mi

12 mi

9. _____

10.   92 ft

43 ft

10. _____

Chapter 1 WHOLE NUMBERS

1.5 Division

> **Learning objectives**
> a   Divide whole numbers.

Key Terms
Use the vocabulary terms listed below to complete the statement in Exercise 1.

quotient   **dividend**   **divisor**

1.   We divide the _____ by the _____ to get the

_____ .

Objective c   Divide whole numbers.

Divide, if possible. If not possible, write "not defined."

2. $63 \div 7$ 2._____

3. $\dfrac{14}{14}$ 3._____

4. $32 \div 0$ 4._____

5. $483 \div 6$ 5._____

6. $4\overline{)7632}$ 6._____

7. $30\overline{)785}$

8. $27\overline{)5943}$

9. $692 \div 3$

10. $56,000 \div 100$

11. $5\overline{)8138}$

12. $32\overline{)9824}$

7. _____

8. _____

9. _____

10. _____

11. _____

12. _____

10

Chapter 1 WHOLE NUMBERS

1.6 Rounding and Estimating; Order

> **Learning Objectives**
> a   Round to the nearest ten, hundred, or thousand.
> b   Estimate sums, differences, and products by rounding.
> c   Use < or > for □ to write a true sentence in a situation like 6 □ 10.

Key Terms
Use the vocabulary terms listed below to complete each statement in Exercises 1–2.

equation   **inequality**

1.   A(n) _____ is a sentence that contains an equals sign.

2.   A sentence like 5 > 2 is called a(n) _____ .

Objective a   Round to the nearest ten, hundred, or thousand.

Round 685,147 to the nearest:

3.   ten.   3. _____

4.   hundred.   4. _____

5.   thousand.   5. _____

Round 505,482 to the nearest:

6.   ten.   6. _____

7.   hundred.   7. _____

8.   thousand.   8. _____

Objective b   Estimate sums, differences, and products by rounding.

Estimate the sum or difference by first rounding to the nearest ten. Show your work.

9.   3 8
        + 8 4

9. _____

10.   9 0 2 1
        − 3 8 7 6

10. _____

11. Estimate the difference by first rounding to the nearest thousand. Show your work.

29,024 − 15,483

11. _____

12. Estimate the product by first rounding to the nearest hundred. Show your work.

315 × 1873

12. _____

13. A school takes 228 students to see a musical. The tickets cost $12 each. Estimate the cost of bringing the students by rounding the cost of a ticket and the number of students to the nearest ten.

13. _____

Objective c   Use < or > for ☐ to write a true sentence in a situation like 6 ☐ 10.

14. 21 ☐ 0   14. _____

15. 58 ☐ 73   15. _____

16. 246 ☐ 264   16. _____

17. 130 ☐ 38   17. _____

18. 2008 ☐ 2003   18. _____

12

Chapter 1 WHOLE NUMBERS

1.7 Solving Equations

| **Learning Objectives** |
| a   Solve simple equations by trial. |
| b   Solve equations like $t + 28 = 54$, $28 \cdot x = 168$, and $98 \cdot 2 = y$. |

Key Terms
Use the vocabulary terms listed below to complete each statement in Exercises 1–2.

variable   **solution**

1. A _____ can represent any number.

2. A _____ of an equation is a number that makes the equation true.

Objective a   Solve simple equations by trial.

Solve by trial.

3. $0 + n = 18$ 3. _____

4. $48 \div x = 6$ 4. _____

5. $y - 5 = 9$ 5. _____

6. $13 \cdot y = 13$ 6. _____

Objective b   Solve equations like $t + 28 = 54$, $28 \cdot x = 168$, and $98 \cdot 2 = y$.

Solve. Be sure to check.

7. $16 + x = 84$ 7. _____

8. $23 = y + 23$ 8. _____

9. $6 \cdot m = 54$ 9. _____

10. $108 = n \cdot 4$

11. $52 \cdot 36 = x$

12. $p = 136 \div 8$

13. $r = 914 + 385$

14. $43,295 + 18,088 = y$

15. $8 \cdot x = 96$

16. $450 = 3 \cdot w$

17. $12 + m = 80$

18. $43 = 28 + x$

19. $5 \cdot z = 3985$

20. $y + 62 = 128$

21. $391 = 17 \cdot w$

22. $324 + x = 556$

23. $30 \cdot a = 5400$

24. $8926 = x + 6541$

25. $46 \cdot y = 2714$

10. _____

11. _____

12. _____

13. _____

14. _____

15. _____

16. _____

17. _____

18. _____

19. _____

20. _____

21. _____

22. _____

23. _____

24. _____

25. _____

14

Chapter 1 WHOLE NUMBERS

1.8 Applications and Problem Solving

Learning Objectives
a   Solve applied problems involving addition, subtraction, multiplication, or division of whole numbers.

Key Terms
Use the vocabulary terms listed below to complete each statement in Exercises 1–2.

familiarize   translate

1. Before we can solve a problem mathematically, we must often _____ the situation to an equation.

2. We _____ ourselves with a situation by making sure we understand what we are being asked to find.

Objective a   Solve applied problems involving addition, subtraction, multiplication, or division of whole numbers.

Solve.

3. Kevin took 24 credits his first year, 12 credits his second year, 18 credits his third year, and 24 credits his fourth year. How many credits did he take in all four years?

    3. _____

4. Refer to the information in Exercise 3. How many more credits did Kevin take in his fourth year than in his second year?

    4. _____

5. A gardener plants four rows of peas with 15 pea plants in each row. How many pea plants are there altogether?

    5. _____

6. Ben bakes 72 cookies in 6 equal batches. How many cookies are in each batch?

6. _____

7. Gavin got a speeding ticket for traveling at 78 mph in a 65 mph zone. How many miles per hour over the 65 mph speed limit was he traveling?

7. _____

8. Barbara read 84 pages while her daughter read 12 pages. How many times the number of pages did Barbara read than her daughter read?

8. _____

9. There are 8 ounces in a cup and 16 cups in a gallon. How many ounces are in a gallon?

9. _____

10. Brendan borrowed $7800 for a four wheeler. If the loan is to be paid off in 60 equal payments, how much is each payment (excluding interest)?

10. _____

11. The distance from Chester to Bedford is 235 miles. The distance from Rochester to Bedford is 1119 miles. How much farther is Bedford from Rochester than from Chester?

11. _____

12. Calliope wrote checks for $26, $334, $78, and $159 on   12. _____
    Saturday. Find the total amount she wrote in checks that
    day.

13. A rectangular room is tiled with 24 tiles across its length   13. _____
    and 18 tiles across its width. How many tiles are there in
    the room?

14. Jon buys three tee shirts for $7 each and pays for them   14. _____
    with a $50 bill. How much change should he receive
    back?

15. The balance in your bank account is $987. You write   15. _____
    checks for $38 and $295 and then deposit a check for $19.
    What is your new balance?

16. Geneva's population was 1,038,274. Five years later the   16. _____
    population was 1,206,436. How much did the population
    increase?

17. Bette's car gets 24 miles to the gallon (mpg). How many   17. _____
    gallons should she expect to use in a 600 mile trip?

18. Jeffrey bought a stereo for $350, a set of headphones for $24, and a CD for $16. How much did he spend altogether?

18. _____

19. Devon donates $15 each to six charities. How much does she donate altogether?

19. _____

20. A package of wrapping paper contains 6 sheets of paper that are 24 in. by 36 in. Find the total area of the paper in the package.

20. _____

21. Jordan treated himself and three of his friends to lunch. They each ordered the lunch special for $6. Jordan paid with two $20 bills. How much change should he receive?

21. _____

22. A rectangular field has a length of 85 feet and a width of 60 feet. Find its area and its perimeter.

22. _____

Chapter 1 WHOLE NUMBERS

1.9 Exponential Notation and Order of Operations

Learning Objectives
a   Write exponential notation for products such as $4 \cdot 4 \cdot 4$.
b   Evaluate exponential notation.
c   Simplify expressions using rules for order of operations.
d   Remove parentheses within parentheses.

Key Terms
Use the vocabulary terms listed below to complete each statement in Exercises 1–4.

    exponent   **base**   **squared**   **cubed**

1.   In the expression 4^5, the 4 is the _____ .

2.   If a number is _____, the exponent in the expression is three.

3.   In the expression 10^4, the 4 is the _____ .

4.   If a number is _____ , the exponent in the expression is two.

Objective a   Write exponential notation for products such as $4 \cdot 4 \cdot 4$.

Write exponential notation.

5. $6 \cdot 6 \cdot 6$ 5. _____

6. $8 \cdot 8 \cdot 8 \cdot 8 \cdot 8$ 6. _____

7. $11 \cdot 11 \cdot 11 \cdot 11$ 7. _____

8. $9 \cdot 9$ 8. _____

Objective b   Evaluate exponential notation.

Evaluate.

9. 3^2

10. 10^6

11. 6^4

12. 13^3

9. _____

10. _____

11. _____

12. _____

Objective c   Simplify expressions using rules for order of operations.

Simplify.

13. $(9+5)+6$

14. $80-(13+4)$

15. $(25-11)-6$

16. $120 \div (20 \div 4)$

17. $(4+3)^3$

18. $6^2 + 9^2$

13. _____

14. _____

15. _____

16. _____

17. _____

18. _____

20

19. $(36-24)^2 - (16-8)^2$ **19.** _____

20. $5 \cdot 8 - 6$ **20.** _____

21. $240 \div 30 + 10$ **21.** _____

22. $5(12-4)^2 - 3(1+6)^2$ **22.** _____

23. $8^2 - 4^3 \div 4^2$ **23.** _____

24. $10^2 - 3 \cdot 8 - (16 + 2 \cdot 3)$ **24.** _____

25. Find the average of 16, 24, and 32 .   **25.** _____

Objective d   Remove parentheses within parentheses.

Simplify.

26. $4 \times 6 + \left\{ 36 \div \left[20 - (12 + 4) \right] \right\}$ **26.** _____

27. $75 \div 5 - \left\{ 3 \times \left[14 - (3 + 6) \right] \right\}$ **27.** _____

28. $\left[40-(2+5)+1\right]-\left[12\div(6\div2)\right]$

28. _____

29. $5\times\left\{(300-30\div5)-\left[(24\div4)-8\div2\right]\right\}$

29. _____

30. $\left(32-2^4\right)^2\div\left(8^2\div2^2\right)^2$

30. _____

Chapter 2 FRACTION NOTATION: MULTIPLICATION AND DIVISION

2.1 Factorizations

Learning Objectives
a   Determine whether one number is a factor of another, and find the factors of a number.
b   Find some multiples of a number, and determine whether a number is divisible by another.
c   Given a number from 1 to 100, tell whether it is prime, composite, or neither.
d   Find the prime factorization of a composite number.

Key Terms
Use the vocabulary terms listed below to complete each statement in Exercises 1–4.

   multiple   **divisible**   **prime**   **composite**

1. A number is a _____ of another if it is the product of the number and

    some natural number.

2. A _____ number has exactly two different factors.

3. A number *a* is _____ by another number *b*   if there is a number *c*

    such that $b \cdot c = a$.

4. A _____ number has more than two different factors.

Objective a   Determine whether one number is a factor of another, and find the factors of a number.

Determine whether the second number is a factor of the first.

5. 45; 5   5. _____

6. 571; 3   6. _____

Find all the factors of the number.

7. 12

7. _____

8. 29

8. _____

9. 96

9. _____

Objective b Find some multiples of a number, and determine whether a number is divisible by another.

Multiply by 1, 2, 3, and so on, to find 10 multiples of the number.

10. 6

10. _____

11. 15

11. _____

Determine whether the first number is divisible by the second number.

12. 42; 6

12. _____

13. 936; 4

13. _____

14. 13,807; 7

14. _____

Objective c   Given a number from 1 to 100, tell whether it is prime, composite, or neither.

Determine whether the number is prime, composite, or neither.

15. 24   15. _____

16. 1   16. _____

17. 73   17. _____

18. 57   18. _____

19. 81   19. _____

Objective d   Find the prime factorization of a composite number.

Find the prime factorization of the number.

20. 21   20. _____

21. 120   21. _____

22. 98   22. _____

23. 45

23. _____

24. 1500

24. _____

25. 4420

25. _____

26

Chapter 2 FRACTION NOTATION: MULTIPLICATION AND DIVISION

2.2 Divisibility

Learning Objectives

a   Determine whether a number is divisible by 2, 3, 4, 5, 6, 7, 8, 9, or 10.

Key Terms

Use the vocabulary terms listed below to complete each statement in Exercises 1–4.

| **divisible** | **divisible by 3** | **divisible by 5** | **even** |

1. A number is divisible by two if it is _____ .

2. If a number has a ones digit of 0 or 5, then it is _____ .

3. A number is _____ if the sum of its digits is divisible by three.

4. A number a is _____ by another number b if there is a number c

   such that $a = b \cdot c$.

Objective a   Determine whether a number is divisible by 2, 3, 4, 5, 6, 7, 8, 9, or 10.

Consider the following numbers for Exercises 5–13.

48	65	41	627
130	26	87	900
264	352	1015	13,956
12,807	599	17	111

5. Which of the above are divisible by 2?   5. _____

6. Which of the above are divisible by 3?   6. _____

7. Which of the above are divisible by 4?   7. _____

8. Which of the above are divisible by 5?   8. _____

9. Which of the above are divisible by 6?   9. _____

10. Which of the above are divisible by 7?   10. _____

11. Which of the above are divisible by 8?   11. _____

12. Which of the above are divisible by 9?   12. _____

13. Which of the above are divisible by 10?   13. _____

Consider the following numbers for Exercises 14–22.

25	17	47	514
78	54	112	369
351	211	980	2039
19,066	3054	8888	15,000

14. Which of the above are divisible by 2?   14. _____

15. Which of the above are divisible by 3?   15. _____

16. Which of the above are divisible by 4?   16. _____

17. Which of the above are divisible by 5?   17. _____

18. Which of the above are divisible by 6?   18. _____

19. Which of the above are divisible by 7?   19. _____

20. Which of the above are divisible by 8?   20. _____

21. Which of the above are divisible by 9?   21. _____

22. Which of the above are divisible by 10?   22. _____

Chapter 2 FRACTION NOTATION: MULTIPLICATION AND DIVISION

2.3 Fractions and Fraction Notation

Learning Objectives
a   Identify the numerator and the denominator of a fraction and write fraction notation for part of an object.
b   Simplify fraction notation like $\dfrac{n}{n}$ to 1, $\dfrac{0}{n}$ to 0, and $\dfrac{n}{1}$ to n.

Key Terms

Use the vocabulary terms listed below to complete each statement in Exercises 1–2.

numerator   **denominator**

1. In a fraction, the _____ tell us the unit into which we are partitioning an object.

2. The _____ of a fraction tells us the number of equal parts we are considering.

Objective a   Identify the numerator and the denominator of a fraction and write fraction notation for part of an object.

Identify the numerator and the denominator.

3. $\dfrac{4}{5}$ 3. _____

4. $\dfrac{0}{3}$ 4. _____

5. $\dfrac{12}{7}$ 5. _____

What part of the object is shaded?

6.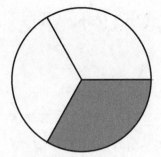

6. _____

7.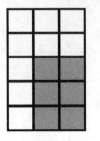

7. _____

Find the ratios.

8. Dan delivers flowers for a florist. On Valentine's Day, he had 50 deliveries scheduled. By 11AM he had delivered 21 orders. What is the ratio of

a. orders delivered to the total number of orders?

8.

a. _____

b. orders delivered to orders not delivered?

b. _____

c. orders not delivered to the total number of orders?

c. _____

30

9. Ruben sells cotton candy at a baseball game. He had 100   9.
    bags to sell. At the seventh inning stretch he had sold 79
    bags. What is the ratio of
    a. bags sold to the total number of bags?   a. _____

    b. bags sold to bags not sold?   b. _____

    c. bags not sold to the total number of bags?   c. _____

Objective b   Simplify fraction notation like $\dfrac{n}{n}$ to 1, $\dfrac{0}{n}$ to 0, and $\dfrac{n}{1}$ to n.

Simplify.

10. $\dfrac{4}{1}$ 10. _____

11. $\dfrac{0}{6}$ 11. _____

12. $\dfrac{15}{15}$ 12. _____

13. $\dfrac{9}{0}$ 13. _____

31

14. $\dfrac{8-4}{3-2}$

14. _____

15. $\dfrac{6}{12-12}$

15. _____

Chapter 2 FRACTION NOTATION: MULTIPLICATION AND DIVISION

2.4 Multiplication and Applications

Learning Objectives
a   Multiply a fraction by a fraction, and multiply a fraction by a whole number.
b   Solve applied problems involving multiplication of fractions.

Key Terms

Use the vocabulary terms listed below to complete each statement in Exercises 1–3.

    area   **solve**   **check**

1.   To _____ an equation means to find all values of a variable that make the equation true.

2.   The _____ of a rectangle is length times width.

3.   We can sometimes _____ a solution by repeating calculations, finding the solution in a different manner, or seeing if the solution is reasonable.

Objective a   Multiply a fraction by a fraction, and multiply a fraction by a whole number.

Multiply.

4. $\dfrac{1}{5} \cdot \dfrac{1}{3}$ 4. _____

5. $\dfrac{3}{8} \cdot \dfrac{3}{8}$ 5. _____

6. $\dfrac{5}{6} \cdot \dfrac{7}{8}$ 6. _____

7. $\dfrac{4}{9} \times \dfrac{1}{3}$

7. _____

8. $6 \times \dfrac{3}{5}$

8. _____

9. $\dfrac{2}{5} \cdot 4$

9. _____

10. $5 \times \dfrac{1}{6}$

10. _____

Objective b Solve applied problems involving multiplication of fractions.

Solve.

11. If each bag contains $\dfrac{5}{8}$ lb of seed, how much does half a bag contain?

11. _____

12. If it takes $\dfrac{3}{4}$ cup sugar to make one batch of cookies, how much does it take to make 3 batches?

12. _____

34

13. How many $\frac{5}{8}$ in. wide books can be placed on a shelf 12 in. wide?

13. _____

14. A water bottle can hold $\frac{5}{8}$ cup. How much will the water bottle hold when it is $\frac{1}{2}$ full?

14. _____

15. It takes $\frac{1}{2}$ yard of dental floss for each patient. How many yards of floss are needed by a dental hygienist for 9 patients?

15. _____

16. A rectangular table top measures $\frac{5}{8}$ m long by $\frac{3}{8}$ m wide. What is the area?

16. _____

17. The recipe for macaroni salad calls for $\frac{1}{3}$ cup of mayonnaise. How much is needed to make $\frac{1}{2}$ of the recipe?

17. _____

18. Seven of 53 high school band members play in a college band. Three of 79 college band members play in a community band. What fractional part of high school band members play in a community band?

18. _____

Chapter 2 FRACTION NOTATION: MULTIPLICATION AND DIVISION

2.5 Simplifying

Learning Objectives
a   Multiply a number by 1 to find fraction notation with a specified denominator.
b   Simplify fraction notation.
c   Use the test for equality to determine whether two fractions name the same number.

Key Terms
Use the vocabulary terms listed below to complete each statement in Exercises 1–3.

canceling   **equivalent**   **simplest**

1. _____ is a shortcut for removing a factor of 1, which must be used with caution.

2. Two fractions are _____ if they name the same number.

3. A fraction is in _____ form if the numerator and the denominator are the smallest whole numbers possible.

Objective a   Multiply a number by 1 to find fraction notation with a specified denominator.

Find another name for the given number, but with the denominator indicated. Use multiplying by 1.

4. $\dfrac{1}{3} = \dfrac{?}{12}$ 4. _____

5. $\dfrac{5}{6} = \dfrac{?}{30}$ 5. _____

6. $\dfrac{8}{3} = \dfrac{?}{24}$ 6. _____

7. $\dfrac{11}{15} = \dfrac{?}{75}$ 7. _____

Objective b   Simplify fraction notation.

Simplify.

8. $\dfrac{9}{12}$

9. $\dfrac{15}{3}$

10. $\dfrac{8}{32}$

11. $\dfrac{18}{24}$

12. $\dfrac{35}{84}$

8. _____

9. _____

10. _____

11. _____

12. _____

Objective c   Use the test for equality to determine whether two fractions name the same number.

Use = or ≠ for ☐ to write a true sentence.

13. $\dfrac{3}{5}$ ☐ $\dfrac{12}{20}$

14. $\dfrac{4}{9}$ ☐ $\dfrac{32}{56}$

15. $\dfrac{8}{15}$ ☐ $\dfrac{3}{5}$

16. $\dfrac{15}{20}$ ☐ $\dfrac{3}{4}$

17. $\dfrac{3}{2}$ ☐ $\dfrac{12}{9}$

18. $\dfrac{7}{8}$ ☐ $\dfrac{42}{48}$

13. _____

14. _____

15. _____

16. _____

17. _____

18. _____

38

Chapter 2 FRACTION NOTATION: MULTIPLICATION AND DIVISION

2.6 Multiplying, Simplifying, and Applications

Learning Objectives
a   Multiply and simplify using fraction notation.
b   Solve applied problems involving multiplication of fractions.

Objective a   Multiply and simplify using fraction notation.

Multiply and simplify.

1. $\dfrac{3}{5} \cdot \dfrac{1}{3}$

1. _____

2. $\dfrac{1}{6} \cdot \dfrac{3}{8}$

2. _____

3. $\dfrac{9}{20} \cdot \dfrac{5}{3}$

3. _____

4. $8 \cdot \dfrac{1}{8}$

4. _____

5. $\dfrac{6}{7} \cdot \dfrac{7}{6}$

5. _____

6. $\dfrac{12}{25} \cdot \dfrac{15}{32}$

6. _____

Objective b   Solve applied problems involving multiplication of fractions.

Solve.

7. Jackson's tuition was $5200. A loan was obtained for $\frac{7}{8}$ of the tuition. How much was the loan?

7. _____

8. A house worth $297,000 is assessed for $\frac{2}{3}$ of its value. What is the assessed value of the home?

8. _____

9. A recipe for pizza crust calls for $\frac{1}{3}$ cup of oil. Juan is making $\frac{1}{2}$ of the recipe. How much oil should he use?

9. _____

10. Telemarketers have determined that $\frac{1}{3}$ of phone numbers will change in 2 years. A business has a phone list of 3510 people. After two years, how many phone numbers on the list will be incorrect?

10. _____

11. On a map, 1 in. represents 180 mi. What distance does $\frac{3}{4}$ in. represent?

11. _____

12. The pitch of a screw is $\frac{5}{32}$ in. How far will it go into a piece of oak when it is turned 6 complete rotations clockwise?

12. _____

40

Chapter 2 FRACTION NOTATION: MULTIPLICATION AND DIVISION

2.7 Division and Applications

Learning Objectives
a   Find the reciprocal of a number.
b   Divide and simplify using fraction notation.
c   Solve equations of the type $a \cdot x = b$ and $x \cdot a = b$, where a and b may be fractions.
d   Solve applied problems involving division of fractions.

Key Terms
Use the vocabulary terms listed below to complete each statement in Exercises 1–2. Terms may be used more than once.

reciprocal $\dfrac{n}{0}$

1. The number zero has no _____ because _____ is not defined.

2. We find the _____ of a fraction by interchanging the numerator and the denominator.

Objective a   Find the reciprocal of a number.

Find the reciprocal of each number.

3. $\dfrac{3}{4}$ 3. _____

4. 9   4. _____

5. $\dfrac{1}{5}$ 5. _____

6. $\dfrac{13}{7}$ 6. _____

7. $\dfrac{5}{8}$ 7. _____

8. $\dfrac{14}{17}$

8. _____

Objective b   Divide and simplify using fraction notation.

Divide and simplify.

9. $\dfrac{5}{6} \div \dfrac{5}{11}$

9. _____

10. $\dfrac{4}{9} \div \dfrac{9}{20}$

10. _____

11. $\dfrac{3}{5} \div \dfrac{6}{7}$

11. _____

12. $\dfrac{9}{5} \div \dfrac{1}{5}$

12. _____

13. $\dfrac{15}{4} \div 3$

13. _____

14. $30 \div \dfrac{5}{6}$

14. _____

15. $\dfrac{2}{3} \div \dfrac{2}{3}$

15. _____

Objective c   Solve equations of the type $a \cdot x = b$ **and** $x \cdot a = b$**, where** a **and** b **may be fractions.**

Solve.

16. $\dfrac{2}{3} \cdot x = 42$

16. _____

17. $\dfrac{5}{7} \cdot y = \dfrac{15}{7}$

17. _____

18. $\dfrac{6}{11} \cdot m = \dfrac{12}{55}$

18. _____

19. $n \cdot \dfrac{5}{4} = 120$

19. _____

20. $p \cdot \dfrac{8}{15} = \dfrac{4}{45}$

20. _____

21. $x \cdot \dfrac{3}{10} = \dfrac{9}{100}$

21. _____

Objective d   Solve applied problems involving division of fractions.

Solve.

22. A pair of children's shorts requires $\frac{2}{3}$ yd of cotton. How many pairs of shorts can be made from 24 yd of cotton?

22.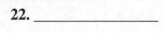

23. How many $\frac{3}{4}$-cup sugar bowls can be filled from 15 cups of sugar?

23.

24. A bucket had 14 L of water in it when it was $\frac{2}{3}$ full. How much could it hold when full?

24. _____

25. A tank had 20 L of gasoline in it when it was $\frac{5}{6}$ full. How much could it hold when full?

25. _____

26. Matthew Le is going to a Penn State football game. After driving 192 kilometers (km), he has completed $\frac{3}{5}$ of the trip. How long is the total trip? How many kilometers are left to drive?

26. _____

27. A piece of cable $\frac{7}{8}$ meters (m) long is to be cut into 14 pieces of the same length. What is the length of each piece?

27. _____

44

Chapter 3 FRACTION NOTATION AND MIXED NUMERALS

3.1 Least Common Multiples

Learning Objectives
a   Find the least common multiple, or LCM, of two or more numbers.

Key Terms
Use the vocabulary terms listed below to complete each statement in Exercises 1–3.

least common multiple	factorization
prime factorization	multiple

1. We find the _____ of a number when we write it as a product of

    primes.

2. $3 \cdot 4$ is an example of a _____ of 12.

3. The _____ of the numbers is the smallest number that is a

    _____ of both numbers.

Object a   Find the least common multiple, LCM, of two or more numbers.

Find the LCM of the set of numbers.

4. 3, 6   4. _____

5. 6, 9   5. _____

6. 20, 50   6. _____

7. 24, 28   7. _____

8. 25, 60   8. _____

9. 12, 44

10. 5, 8, 16

11. 15, 35

12. 9, 49

13. 11, 17

14. 24, 54

15. 6, 33

16. 2, 5, 7

17. 16, 36

18. 4, 9, 12

19. 15, 8, 30, 12

20. 33, 27, 54

21. 75, 125, 50, 500

9. _____

10. _____

11. _____

12. _____

13. _____

14. _____

15. _____

16. _____

17. _____

18. _____

19. _____

20. _____

21. _____

Josie takes 5 min to run a lap, Bob takes 6 min to run a lap, and Karen takes 8 min to run a lap. All three began running laps at the same location at the same time. Use this information for Exercises 22-25.

22. When will Josie and Bob again meet at the starting place?

22. _____

23. When will Bob and Karen again meet at the starting place?

23. _____

24. When will Josie, Bob, and Karen again meet at the starting place?

24. _____

Chapter 3 FRACTION NOTATION AND MIXED NUMERALS

3.2 Addition and Applications

Learning Objectives
a   Add using fraction notation.
b   Solve applied problems involving addition with fraction notation.

Key Terms
Use the vocabulary terms listed below to complete each statement in Exercises 1–3.

> **least common denominator**   **like denominators**
> **different denominators**

1.   We find the LCD before adding or subtracting fractions which have

    _____ .

2.   The _____ of two fractions is the LCM of the two denominators.

3.   When adding fractions with _____ , we add the numerators, keep the

    denominator, and simplify, if possible.

Objective a   Add using fraction notation.

Add and simplify.

4. $\dfrac{3}{4} + \dfrac{1}{4}$ 4. _____

5. $\dfrac{1}{8} + \dfrac{5}{8}$ 5. _____

6. $\dfrac{9}{10} + \dfrac{3}{5}$ 6. _____

7. $\dfrac{4}{15} + \dfrac{5}{6}$ 7. _____

8. $\dfrac{3}{10} + \dfrac{21}{100}$

8. _____

9. $\dfrac{3}{8} + \dfrac{5}{12} + \dfrac{1}{3}$

9. _____

10. $\dfrac{7}{9} + \dfrac{16}{45} + \dfrac{2}{25}$

10. _____

Objective b   Solve applied problems involving addition with fraction notation.

11. Danielle bought $\dfrac{2}{3}$ lb of green tea and $\dfrac{3}{4}$ lb of English Breakfast tea. How many pounds of tea did she buy?

11. _____

12. A recipe for punch calls for $\dfrac{1}{4}$ qt of ginger ale and $\dfrac{2}{3}$ qt of orange juice. How much liquid is needed? If the recipe is doubled, how much liquid is needed? If the recipe is halved, how much liquid is needed?

12. _____

13. A baker used $\dfrac{1}{3}$ lb sugar for candy, $\dfrac{1}{4}$ lb sugar for cookies and $\dfrac{1}{8}$ lb sugar for muffins. How much sugar was used?

13. _____

14. To make a small meatloaf, Jody bought $\dfrac{7}{8}$ lb ground beef, $\dfrac{1}{3}$ lb ground pork, and $\dfrac{1}{4}$ lb ground veal. How much ground meat was there?

14. _____

15. A $\dfrac{11}{16}$ in. thick tile is glued to a board that is $\dfrac{15}{16}$ in. thick. The glue is $\dfrac{3}{32}$ in. thick. How thick is the result?

15. _____

48

Chapter 3 FRACTION NOTATION AND MIXED NUMERALS

3.3 Subtraction, Order, and Applications

Learning Objectives
a   Subtract using fraction notation.
b   Use < or > with fraction notation to write a true sentence.
c   Solve equations of the type $x + a = b$ and $a + x = b$, where a and b may be fractions.
d   Solve applied problems involving subtraction with fraction notation.

Objective a   Subtract using fraction notation.

Subtract and simplify.

1. $\dfrac{7}{8} - \dfrac{3}{8}$

1. _____

2. $\dfrac{9}{10} - \dfrac{3}{5}$

2. _____

3. $\dfrac{11}{16} - \dfrac{5}{12}$

3. _____

4. $\dfrac{73}{100} - \dfrac{3}{10}$

4. _____

5. $\dfrac{4}{5} - \dfrac{2}{3}$

5. _____

6. $\dfrac{5}{2} - \dfrac{13}{24}$

6. _____

Objective b   Use < or > with fraction notation to write a true sentence.

Use < or > for □ to write a true sentence.

7. $\dfrac{3}{7} \ \square \ \dfrac{2}{7}$

7. _____

8. $\dfrac{1}{9} \ \square \ \dfrac{1}{16}$

8. _____

9. $\dfrac{2}{5} \ \square \ \dfrac{4}{9}$

9. _____

10. $\dfrac{13}{4} \ \square \ \dfrac{10}{3}$

10. _____

11. $\dfrac{9}{16} \ \square \ \dfrac{7}{12}$

11. _____

Objective c   Solve equations of the type $x + a = b$ and $a + x = b$, where a and b may be fractions.

Solve.

12. $x + \dfrac{1}{3} = \dfrac{5}{6}$

12. _____

13. $y + \dfrac{4}{9} = \dfrac{7}{9}$

13. _____

14. $\dfrac{5}{6} + z = \dfrac{9}{10}$

14. _____

50

15. $x + \dfrac{2}{13} = \dfrac{5}{8}$

15. _____

16. $y + \dfrac{1}{8} = \dfrac{7}{6}$

16. _____

Objective d   Solve applied problems involving subtraction with fraction notation.

Solve.

17. In preparation for a race, Dee ran $\dfrac{2}{3}$ mile every day. One day she had already run $\dfrac{1}{4}$ mile. How much farther should Dee run?

17. _____

18. An inheritance was left to five cousins. One received $\dfrac{1}{3}$ of the property, the second $\dfrac{1}{8}$, and the third $\dfrac{1}{16}$, and the fourth $\dfrac{3}{8}$. How much did the fifth receive?

18. _____

19. From a $\frac{3}{4}$-lb wheel of cheese, a $\frac{1}{3}$-lb piece was served. How much cheese remained on the wheel?

20. Sarah has an $\frac{11}{8}$-lb mixture of raisins and sunflower seeds that includes $\frac{3}{4}$ lb of raisins. How many pounds of sunflower seeds are in the mixture?

21. Ray has a pitcher containing $\frac{13}{16}$ cup of milk and puts $\frac{1}{12}$ cup in his coffee. How much milk remains in his pitcher?

Chapter 3 FRACTION NOTATION AND MIXED NUMERALS

3.4 Mixed Numerals

Learning Objectives
a Convert between mixed numerals and fraction notation.
b Divide whole numbers, writing the quotient as a mixed numeral.

Key Terms

Use the vocabulary terms listed below to complete each statement in Exercises 1–2. Terms may be used more than once.

    mixed numeral(s)   **fraction notation**

1. To divide using _____, first write _____ and divide. Then convert the answer to a mixed numeral, if appropriate.

2. _____ consist of a whole number part and a fraction less than 1.

Objective a Convert between mixed numerals and fraction notation.

Convert to fraction notation.

3. $4\dfrac{1}{2}$ **3.** _____

4. $5\dfrac{5}{6}$ **4.** _____

5. $25\dfrac{5}{8}$ **5.** _____

Convert to a mixed numeral.

6. $\dfrac{35}{2}$ **6.** _____

7. $\dfrac{23}{10}$ **7.** _____

8. $\dfrac{264}{7}$

8. _____

Objective b   Divide whole numbers, writing the quotient as a mixed numeral.

Divide. Write a mixed numeral for the answer.

9. $7\overline{)963}$

9. _____

10. $4\overline{)4229}$

10. _____

11. $65\overline{)7432}$

11. _____

12. $313 \div (-16)$

12. _____

13. $-509 \div 12$

13. _____

Chapter 3 FRACTION NOTATION AND MIXED NUMERALS

3.5 Addition and Subtraction Using Mixed Numerals; Applications

Learning Objectives
a   Add using mixed numerals.
b   Subtract using mixed numerals.
c   Solve applied problems involving addition and subtraction with mixed numerals.

Objective a   Add using mixed numerals.

Add. Write a mixed numeral for the answer.

1. $3\dfrac{4}{5}$

     $+\,5\dfrac{2}{5}$

1. _____

2. $9\dfrac{2}{3}$

     $+\,6\dfrac{2}{3}$

2. _____

3. $1\dfrac{3}{7}+4\dfrac{1}{2}$

3. _____

4. $6\dfrac{3}{4}+5\dfrac{1}{8}$

4. _____

5. $8\dfrac{5}{12}$

$+\ 7\dfrac{7}{8}$

6. $18\dfrac{1}{3}$

$23\dfrac{7}{8}$

$+\ 39\dfrac{5}{6}$

Objective b   Subtract using mixed numerals.

Subtract. Write a mixed numeral for the answer.

7. $9\dfrac{3}{10}$

$-\ 4\dfrac{7}{10}$

8. $8\dfrac{2}{3}-2\dfrac{1}{8}$

9. 22

$-\ 16\dfrac{5}{6}$

10. $13\dfrac{2}{5}$

     $-\,8\dfrac{3}{4}$

10. _____

11. $36\dfrac{1}{5}$

     $-\quad\dfrac{7}{8}$

11. _____

12. $45\dfrac{2}{3}$

     $-\,9$

12. _____

Objective c   **Solve applied problems involving addition and subtraction with mixed numerals.**

Solve.

13. Lara is 65 in. tall and her son, Tim is $58\dfrac{5}{12}$ in. tall. How

much taller is Lara?

13. _____

14. Michelle bought $1\frac{2}{3}$ lb red delicious apples and $3\frac{4}{5}$ lb of gala apples. What was the total weight of the apples?

15. Renee bought two types of drapery fabric. She needs $9\frac{3}{4}$ yd of the floral print and $10\frac{7}{8}$ yd of solid print. How many yards did Renee buy?

16. Jon had $4\frac{1}{3}$ gal of paint. It took $2\frac{2}{3}$ gal to paint the kitchen. He estimates that it would take $3\frac{1}{4}$ gal to paint the living room. How much more paint did Jon need?

17. A picture frame has dimensions of $28\frac{5}{8}$ in. wide $32\frac{1}{4}$ in. long. What is the perimeter of the framed picture?

58

Chapter 3 FRACTION NOTATION AND MIXED NUMERALS

3.6 Multiplication and Division Using Mixed Numerals; Applications

Learning Objectives
a   Multiply using mixed numerals.
b   Divide using mixed numerals.
c   Solve applied problems involving multiplication and division with mixed numerals.

Objective a   Multiply using mixed numerals.

Multiply. Write a mixed numeral for the answer.

1. $3 \cdot 5\frac{4}{7}$

1. _____

2. $12\frac{3}{4} \cdot \frac{1}{2}$

2. _____

3. $5\frac{7}{10} \cdot 6\frac{2}{3}$

3. _____

4. $8\frac{7}{8} \cdot 3\frac{5}{8}$

4. _____

5. $18\frac{2}{5} \cdot 9\frac{4}{5} \cdot 3\frac{1}{3}$

5. _____

Objective b   Divide using mixed numerals.

Divide. Write a mixed numeral for the answer.

6. $18 \div 6\frac{2}{3}$

6. _____

7. $4\frac{1}{2} \div 3$

7. _____

8. $5\frac{5}{8} \div 2\frac{2}{9}$

8. _____

9. $3\frac{1}{8} \div 1\frac{1}{8}$

9. _____

10. $4\frac{1}{4} \div 16$

10. _____

Objective c   Solve applied problems involving multiplication and division with mixed numerals.

Solve.

11. A mural is $6\frac{5}{8}$ ft by $9\frac{1}{4}$ ft. What is the area of the mural?

11. _____

60

12. The weight of water is $62\frac{1}{2}$ lb per cubic foot. What is the weight of $4\frac{1}{4}$ cubic feet of water?

12.

13. A car traveled 368 mi on $11\frac{1}{2}$ gal of gas. How many miles per gallon did it get?

13.

14. The recommended serving of salmon is $\frac{1}{4}$ lb. How many servings can be prepared from $4\frac{1}{3}$ lb of salmon?

14.

15. Fahrenheit temperature can be obtained from Celsius (Centigrade) temperature by multiplying by $1\frac{4}{5}$ and adding 32°. What Fahrenheit temperature corresponds to a Celsius temperature of 25°?

15.

16. A walkway is to be $15\frac{3}{4}$ yd long. Rectangular stone tiles that are each $\frac{7}{8}$ yd long are used to form the walk. How many tiles are used?

16. _____

17. A cinnamon roll contains $1\frac{1}{3}$ tsp of cinnamon. How much cinnamon is need for 20 rolls?

18. A rectangular lot has dimensions of $305\frac{1}{2}$ ft by $211\frac{1}{4}$ ft. A building with dimensions 80 ft by $30\frac{1}{2}$ ft is built on the lot. How much area is left over?

19. The weight of water is $8\frac{1}{3}$ lb per gallon. Tyler rolls his lawn with an 850-lb capacity roller. Express the water capacity of the roller in gallons.

20. Most space shuttles orbit Earth once every $1\frac{1}{2}$ hr. How many orbits are made every week?

Chapter 3 FRACTION NOTATION AND MIXED NUMERALS

3.7 Order of Operations; Estimation

Learning Objectives
a   Simplify expressions using the rules for order of operations.
b   Estimate with fraction notation and mixed numerals.

Key Terms
Use the vocabulary terms listed below to complete each statement in Exercises 1–2.

    simplify   **order of operations**   **average**

1.   To compute a(n) _____ we find the sum of the numbers, then divide by the number of addends.

2.   The rules for _____ allow us to _____ an expression by indicating which calculation to perform first, second, and so on to find an equivalent expression.

Objective a   Simplify expressions using the rules for order of operations.

Simplify.

3. $\dfrac{1}{5} \cdot \dfrac{1}{6} \cdot \dfrac{1}{7}$ 3. _____

4. $8 \div 4 \div 3$ 4. _____

5. $\dfrac{2}{3} \div \dfrac{5}{8} \div \dfrac{4}{5}$ 5. _____

6. $\dfrac{7}{12} \div \dfrac{1}{6} - \dfrac{3}{4} \cdot \dfrac{12}{5}$ 6. _____

7. $\dfrac{9}{8} \cdot \dfrac{4}{5} + \dfrac{3}{5} \div 6$

8. $\dfrac{5}{9} - \dfrac{1}{3}\left(\dfrac{1}{6} + \dfrac{5}{9}\right)$

9. $38\dfrac{3}{7} - 8\dfrac{1}{4} + 2\dfrac{1}{2}$

10. $16\dfrac{2}{3} - 3\dfrac{4}{5} - 6\dfrac{7}{8}$

11. $\dfrac{5}{4} \div \dfrac{3}{8} \cdot \dfrac{1}{9}$

12. $\left(\dfrac{2}{3}\right)^2 - \dfrac{1}{21} \cdot 2\dfrac{5}{8}$

13. Find the average of $\dfrac{5}{12}$ and $\dfrac{3}{16}$.

14. Find the average of $4\frac{1}{2}$, $3\frac{3}{4}$, and $5\frac{1}{8}$.

14. _____

Simplify.

15. $\left(\dfrac{3}{8}+\dfrac{7}{6}\right)\div\left(\dfrac{1}{4}-\dfrac{2}{11}\right)$

15. _____

16. $\left(4\dfrac{1}{2}-2\dfrac{3}{5}\right)^2+9\cdot6\dfrac{2}{3}\div4$

16. _____

Objective b   Estimate with fraction notation and mixed numerals.

Estimate each of the following as 0, $\dfrac{1}{2}$, or 1.

17. $\dfrac{3}{50}$

17. _____

18. $\dfrac{15}{17}$

18. _____

19. $\dfrac{11}{25}$

19. _____

20. $\dfrac{37}{35}$

20. _____

Find a number for the blank so that the fraction is close to but greater than $\dfrac{1}{2}$. Answers may vary.

21. $\dfrac{\square}{15}$

21. _____

22. $\dfrac{29}{\square}$

22. _____

65

Find a number for the blank so that the fraction is close to but greater than 1. Answers may vary.

23. $\dfrac{\square}{36}$

24. $\dfrac{9}{\square}$

Estimate each part of the following as a whole number, as $\dfrac{1}{2}$, or as a mixed number where the fractional part is $\dfrac{1}{2}$.

25. $8\dfrac{3}{16}$

26. $4\dfrac{5}{8}$

27. $\dfrac{3}{5} \cdot \dfrac{9}{10}$

28. $4\dfrac{5}{6} \cdot 3\dfrac{1}{10} + 8\dfrac{1}{5}$

29. $18 \div \dfrac{40}{7}$

30. $25\dfrac{3}{4} \cdot 4\dfrac{1}{8} - 5\dfrac{4}{9}$

Chapter 4 DECIMAL NOTATION

4.1 Decimal Notation, Order, and Rounding

Learning Objectives
a   Given decimal notation, write a word name.
b   Convert between decimal notation and fraction notation.
c   Given a pair of numbers in decimal notation, tell which is larger.
d   Round decimal notation to the nearest thousandth, hundredth, tenth, one, ten, hundred, or thousand.

Key Terms
Use the vocabulary terms listed below to complete each statement in Exercises 1–3.

arithmetic number(s)	**decimal point**	**decimal notation**
whole number(s)		**fraction(s)**

1.   _____ consist of _____ and nonnegative

     _____.

2.   An arithmetic number may be written in _____, such as 3.25, or as a

     mixed numeral, such as $3\frac{1}{4}$, or a fraction, such as $\frac{13}{4}$.

3.   When writing a decimal number as a word, we write the word "and" for the

     _____.

Objective a   Given decimal notation, write a word name.

Write a word name for the number in the sentence.

4.   Kevin finished the trial in 348.65 sec.   4. _____

5.   The gasoline costs $2.789 per gallon.   5. _____

6.   A nickel is worth $0.05.   6. _____

Write a word name.

7. 39.073

7. _____

8. 118.3482

8. _____

9. 5.0801

9. _____

Objective b   Convert between decimal notation and fraction notation.

Write fraction notation. Do not simplify.

10. 4.6

10. _____

11. 0.29

11. _____

12. 0.0013

12. _____

13. 205.004

13. _____

Write decimal notation.

14. $\dfrac{6}{10}$

14. _____

15. $\dfrac{4958}{100}$

15. _____

16. $\dfrac{23}{10,000}$

16. _____

17. $5\,\dfrac{81,749}{100,000}$

17. _____

68

Objective c   Given a pair of numbers in decimal notation, tell which is larger.

Which number is larger?

18. 0.03, 0.027   18. _____

19. 0.506, 0.51   19. _____

20. 19.0031, 19.029   20. _____

21. $\dfrac{7}{100}$, 0.007   21. _____

22. 0.916, 0.9164   22. _____

Objective d   Round decimal notation to the nearest thousandth, hundredth, tenth, one, ten, hundred, or thousand.

Round to the nearest tenth.

23. 0.65   23. _____

24. 199.845   24. _____

Round 54,795.06348 to the nearest:

25. thousand.

26. hundred.

27. one.

28. tenth.

29. hundredth.

30. thousandth.

25. _____

26. _____

27. _____

28. _____

29. _____

30. _____

Chapter 4 DECIMAL NOTATION

4.2 Addition and Subtraction

Learning Objectives
a   Add using decimal notation.
b   Subtract using decimal notation.
c   Solve equations of the type $x + a = b$ and $a + x = b$, where a and b may be in decimal notation.
d   Balance a checkbook.

Key Terms

Use the vocabulary terms listed below to complete each statement in Exercises 1–4. Terms may be used more than once.

balance forward   **debit(s)**   **credit(s)**

place-value digits

1. A _____ increases the balance in a checking account.

2. A _____ decreases the balance in a checking account.

3. When we add numbers in decimal notation, we must add corresponding

    _____, so we line up the decimal points.

4. The _____ column in a checkbook indicates how much money is in the

    account after all the _____ and _____ up to that

    point have been calculated.

Objective a   Add using decimal notation.

Add.

5.   381.64
    +   22.93

5. _____

6.   488.307
     +   296.458

6. _____

7.   5.21 + 48

7. _____

8.   4.5 + 0.894 + 54

8. _____

9.   15.74
          8.593
        421.8306
     +   46.912

9. _____

10. 88.5003 + 4765.29 + 300.056 + 490

10. _____

Objective b   Subtract using decimal notation.

Subtract.

11.   42.38
     −   6.54

11. _____

12.   53.471
     −   6.52

12. _____

13.   4.6
     −   0.0035

13. _____

Copyright © 2010 Pearson Education, Inc. Publishing as Addison-Wesley.

Copyright © 2010 Pearson Education, Inc. Publishing as Addison-Wesley.

14. $36.1 - 18.84$

14. _____

15. $8 - 0.29$

15. _____

16. $64 - 5.82$

16. _____

17.
$$\begin{array}{r} 14.0354 \\ -\ \ \ 8.13079 \\ \hline \end{array}$$

17. _____

Objective c   Solve equations of the type $x + a = b$ **and** $a + x = b$, **where** a **and** b **may be in decimal notation.**

Solve.

18. $x + 15.4 = 38.52$

18. _____

19. $y + 0.88 = 115.8$

19. _____

20. $102.05 = t + 64.96$

20. _____

21. $23,009.2 = w + 24.51$

21. _____

22. $z + 0.0054 = 20$

22. _____

23. $403.962 + a = 811.243$

23. _____

Objective d   Balance a checkbook.

24. Tori had $580.66 in her checking account before she wrote checks for payments of $36.43, $290.35, and $119.84, and made a deposit of $254.08. Find her balance after making these transactions.

24. _____

25. Cleo had a balance of $1843.29 in his checking account before he wrote checks for payments of $650.00, $385.15, and $45.99, and made a deposit of $786.35. Find his balance after making these transactions.

25. _____

26. Jerome had a balance of $1209.36 in his checking account before he wrote checks for payments of $286.74 and $314.89, and made deposits of $185.00 and $239.25. Find his balance after making these transactions.

26. _____

27. Polly had a balance of $238.06 in her checking account before she wrote checks for payments of $25.95, $48.79, and $16.30 and made a deposit of $75.00. Find her balance after making these transactions.

27. _____

28. Giles had a balance of $511.53 in his checking account before he wrote checks for payments of $65.00 and $92.14, and made deposits of $47.25 and $31.36. Find his balance after making these transactions.

28. _____

Chapter 4 DECIMAL NOTATION

4.3 Multiplication

Learning Objectives
a Multiply using decimal notation.
b Convert from notation like 45.7 million to standard notation, and convert between dollars and cents.

Key Terms

Use the vocabulary terms listed below to complete each statement in Exercises 1–3. Terms may be used more than once.

cent(s)   **dollar(s)**   **$**   **¢**

1. There are 100 _____ in one _____ .

2. The symbol _____ follows a number to the right and indicates how

   many _____ there are.

3. The _____ symbol precedes a number to the left and indicates how

   many _____ there are.

Objective a   Multiply using decimal notation.

Multiply.

4.   3.4   4. _____
     × ___6

5.   9.2   5. _____
     × _0.7

6.   0.39   6. _____
     × ___4

7.   9.3
         $\times$ 0.06

7. _____

8.   24.6
         $\times$ 0.005

8. _____

9. 10×52.65

9. _____

10. 4.8×1000

10. _____

11. 0.01×489.234

11. _____

12. 0.0653×0.001

12. _____

13.   48.5
          $\times$ 18

13. _____

14.   0.285
          $\times$ 4.4

14. _____

15.   3.574
          $\times$ 2.06

15. _____

16.   4.72
          $\times$ 30

16. _____

76

17.   46.3
         × 10.9   **17.** _____

18.   0.00431
         ×   0.023   **18.** _____

19.   4.02
         × 0.543   **19.** _____

20. 100×204.385   **20.** _____

Objective b   Convert from notation like 45.7 million to standard notation, and convert between dollars and cents.

Convert from dollars to cents.

21. $35.07   **21.** _____

22. $0.72   **22.** _____

23. $2.39   **23.** _____

Convert from cents to dollars.

24. 200¢

24. _____

25. 45¢

25. _____

26. 6¢

26. _____

27. 4862¢

27. _____

Convert the number in the sentence to standard notation.

28. Carter's Window Treatments grossed $2.5 million last year.

28. _____

29. The department store chain had 43.6 billion in sales last year.

29. _____

30. The wave traveled at 0.76 million feet per second.

30. _____

78

Chapter 4 DECIMAL NOTATION

4.4 Division

Learning Objectives
a   Divide using decimal notation.
b   Solve equations of the type $a \cdot x = b,$ where a and b may be in decimal notation.
c   Simplify expressions using the rules for order of operations.

Key Terms
Use the vocabulary terms listed below to complete each statement in Exercises 1–4.

   divisor   **average**   **dividend**   **quotient**

1.   The _____ is the result of a division.

2.   In division, the number being divided is called the _____ .

3.   The _____ of a set of numbers is the sum of the numbers divided by the number of addends.

4.   If division is written in fraction form, such as $\dfrac{483.16}{10}$, the _____ is the denominator of the fraction.

Objective a   Divide using decimal notation.

Divide.

5. $4\overline{)6.48}$ 5. _____

6. $8\overline{)45}$ 6. _____

7. $7\overline{)17.36}$ 7. _____

8. $24\overline{)97.92}$

8. _____

9. $15.6 \div 3$

9. _____

10. $9\overline{)6.3}$

10. _____

11. $0.06\overline{)2.28}$

11. _____

12. $5.6\overline{)168}$

12. _____

13. $3.9\overline{)0.2262}$

13. _____

14. $0.28\overline{)0.1456}$

14. _____

80

15. $0.065 \overline{)5.46}$ **15.** _____

16. $\dfrac{94.2075}{100}$ **16.** _____

17. $\dfrac{3.476}{0.01}$ **17.** _____

Objective b   **Solve equations of the type** $a \cdot x = b,$ **where** a **and** b **may be in decimal notation.**

Solve.

18. $5.4 \cdot x = 38.88$ **18.** _____

19. $14 \cdot y = 13.02$ **19.** _____

20. $100 \cdot z = 5.4702$ **20.** _____

21. $1809.75 = 0.38 \cdot p$ **21.** _____

22. $508 = 20.32 \cdot w$

22. _____

Objective c   Simplify expressions using the rules for order of operations.

Simplify.

23. $15 \times (4.76 + 81.2)$

23. _____

24. $(19.3 - 16.8) \times 21$

24. _____

25. $0.005 + 6.02 \div 0.1$

25. _____

26. $125.6 - 3.75 \times 2.04$

26. _____

27. $24.7 + 3.5 \times (8.4 - 0.25)^2$

27. _____

28. $1.2^2 \div (2 + 4.4) - \left[(3 - 2.6) \times (30 \div 300) \right]$

28. _____

29. $0.05 \times \left\{ 3 \times 48.6 - \left[(12 - 9.2) \div 0.02 \right] \right\}$

29. _____

30. Find the average of \$134.21, \$119.84, \$108.20, and \$156.95.

30. _____

Name: Date:
Instructor: Section:

Chapter 4 DECIMAL NOTATION

4.5 Converting from Fraction Notation to Decimal Notation

Learning Objectives
a   Convert from fraction notation to decimal notation.
b   Round numbers named by repeating decimals in problem solving.
c   Calculate using fraction notation and decimal notation together.

Key Terms
Use the vocabulary terms listed below to complete each statement in Exercises 1–2.

terminating decimal   **repeating decimal**

1.   A _____ occurs if division leads to a repeating pattern of nonzero remainders.

2.   A _____ occurs if division leads to a remainder of zero.

Objective a   Convert from fraction notation to decimal notation.

Find decimal notation.

3. $\dfrac{51}{100}$ 3. _____

4. $\dfrac{7}{100}$ 4. _____

5. $\dfrac{17}{20}$ 5. _____

6. $\dfrac{5}{8}$ 6. _____

7. $\dfrac{31}{25}$

7. _____

8. $\dfrac{17}{16}$

8. _____

9. $\dfrac{3}{7}$

9. _____

10. $\dfrac{5}{6}$

10. _____

11. $\dfrac{17}{12}$

11. _____

12. $\dfrac{5}{11}$

12. _____

Objective b   Round numbers named by repeating decimals in problem solving.

Round the answer to the given exercise to the nearest tenth, hundredth, and thousandth.

13. Exercise 7

13. _____

14. Exercise 8

14. _____

15. Exercise 9

15. _____

16. Exercise 10

16. _____

84

17. Exercise 11   17. _____

18. Exercise 12   18. _____

Round each to the nearest tenth, hundredth, and thousandth.

19. $0.\overline{29}$ 19. _____

20. $4.6\overline{5}$ 20. _____

Find the gas mileage rounded to the nearest tenth.

21. 376 miles; 15 gallons   21. _____

22. 292.8 miles; 14.8 gallons   22. _____

23. Find the average of 27.6, 14.5, 23.2, 24.7, and 22.1   23. _____
    rounded to the nearest tenth.

Objective c   Calculate using fraction notation and decimal notation together.

Calculate.

24. $\dfrac{4}{5} \times 16.625$ 24. _____

25. $8\dfrac{7}{8} - 6.817$

25. _____

26. $\dfrac{3}{4} \times 0.0852 + \dfrac{3}{2} \times 1.408$

26. _____

27. $36.92 \div \dfrac{5}{12} + \dfrac{2}{3} \times 14.4$

27. _____

28. $476.2 \div \dfrac{2}{3} - 15 \times \dfrac{3}{5}$

28. _____

29. $9.6 \times 3\dfrac{1}{2}$

29. _____

30. $5\dfrac{3}{4} \div 4.625$

30. _____

Chapter 4 DECIMAL NOTATION

4.6 Estimating

Learning Objective
a   Estimate sums, differences, products, and quotients.

Key Terms

Use the vocabulary terms listed below to complete each statement in Exercises 1–4.

      addend(s)   **product(s)**   **sum(s)**

      quotient(s)   **difference(s)**

1.   The answer to a subtraction is a(n) _____ .

2.   The numbers we add when finding a(n) _____ are called _____ .

3.   A(n) _____ is found by dividing.

4.   A factor $\times$ a factor is a(n) _____ .

Objective a   Estimate sums, differences, products, and quotients.

Consider the following table for Exercises 5–12. In each exercise, estimate the sum, difference, product, or quotient involved and indicate which of the given choices is an appropriate estimate.

Item	Cost
DVD player	$51.99
CD	$17.95
Portable Stereo	$88.60
Television	$379.00

5.   Estimate the total cost of one DVD player and one   **5.** _____
television.
    a) $4300   b) $420   c) $470   d) $430

6.   Estimate the total cost of one CD and one portable stereo.   **6.** _____
    a) $110   b) $90   c) $70   d)$140

7. About how much more does a television cost than a portable stereo?

    a) $330   b) $2900 c) $290   d) $29

7. _____

8. About how much more does a portable stereo cost than a DVD player?

    a) $50   b) $40   c) $330   d) $400

8. _____

9. Estimate the total cost of 5 CDs.

    a) $100   b) $20   c) $50   d) $125

9. _____

10. Estimate the total cost of 3 televisions.

    a) $11,400   b) $1520 c) $1140 d) $380

10. _____

11. About how many portable stereos can be purchased for $450?

    a) 4   b) 5   c) 1   d) 10

11. _____

12. About how many DVD players can be purchased for $625?

    a) 120   b) 7   c) 10   d) 12

12. _____

Estimate by rounding as directed.

13. $0.3 + 1.45 + 0.54$; nearest tenth

13. _____

14. $0.79 + 3.41 + 5.06$; nearest one

14. _____

15. $2.12 + 3.765 + 84.82$; nearest one

15. _____

16. $16.675 + 3.024$; nearest tenth

16. _____

17. $4.358 - 0.82$; nearest tenth

17. _____

88

18. $38.0543 - 3.76815$; nearest one

18. _____

19. $165.4529 - 38.6147$; nearest ten

19. _____

Estimate. Choose a rounding digit that gives one or two nonzero digits. Indicate which of the choices is an appropriate estimate.

20. $384.3672 - 201.4319$
   a) 18   b) 100   c) 80   d) 180

20. _____

21. 37×5.43
   a) 20   b) 200   c) 2000   d) 150

21. _____

22. 5.3×7.9
   a) 400   b) 4000   c) 40   d) 0.4

22. _____

23. 24.2×0.075
   a) 2   b) 0.02   c) 20   d) 0.2

23. _____

24. 81×3.6
   a) 240   b) 320   c) 24   d) 32

24. _____

25. $4.8 \div 5$
   a) 10   b) 0.1   c) 0.01   d) 1

25. _____

26. $0.09542 \div 1.86$
   a) 0.5   b) 20   c) 0.05   d) 2

26. _____

27. $59.84 \div 21.6$
   a) 3   b) 0.3   c) 30   d) 1200

27. _____

28. $621 \div 0.888$
   a) 6000   b) 60   c) 6   d) 600

28. _____

29. Calvin receives a grant of $700 to purchase technology supplies for his classroom. He decides to purchase graphing calculators which costs $67.86 each. Estimate how many of these calculators he can purchase with his grant money. Show your work.

29. _____

30. Tara wants to paint the outside of her house. The area of her siding is 2356 sq ft. Each gallon of paint will cover 400 sq ft. Estimate how many gallons of paint Tara will need. Show your work.

30. _____

Chapter 4 DECIMAL NOTATION

4.7 Applications and Problem Solving

Learning Objective
a   Solve applied problems involving decimals.

Key Terms
Use the vocabulary terms listed below to complete each statement in Exercises 1–2.

   variable   **estimate**

1.   A(n) _____ is a letter used to represent an unknown quantity.

2.   A partial check of a solution can often be done by finding a(n) _____ .

Objective a   Solve applied problems involving decimals.

Solve.

3.   Garrison's odometer read 58,176.3 before starting his trip.   3. _____
     When he arrived at his destination it read 59,384.1. How
     many miles had he driven?

4.   A group of 5 friends pay $68.55 for lunch. If they share   4. _____
     the cost evenly, how much does each person pay?

5.   Normal body temperature is 98.6°F. During an illness   5. _____
     Grant's temperature was 3.8°F above normal. What was
     his temperature?

Find in the area of the figure.

6.   3.2 cm
5.4 cm

6. _____

7.   10.24 mi
10.24 mi

7. _____

Find the perimeter of the figure.

8.

5.1 in.   4.0 in.
0.9 in.   1.8 in.
4.6 in.

8. _____

9.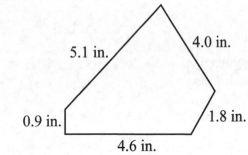

9.75 m
4.5 m
9 m
6.5 m

9. _____

92

Find the length d in the figure.

10. 1.24 cm   1.24 cm

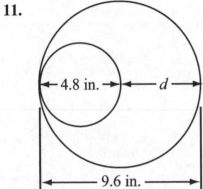

5.76 cm

10. _____

11.

4.8 in.   d

9.6 in.

11. _____

12. Sandy filled her vehicle with gas and noted that the
odometer read 98,107.4. After the next filling, the
odometer read 98,467.9. It took 13.5 gallons to fill the
tank. How many miles per gallon did she get? Round to
the nearest tenth of a mile per gallon.

12. _____

13. The Callahans own a house with an assessed value of $178,300. For every $1000 of assessed value, they pay $4.36 in taxes. How much do they pay in taxes?

13. _____

14. Quinn drives 12.2 miles at 40 mph. How long does she travel? (Use the formula *Distance = Speed × Time*.)

14. _____

15. Lot A measures 50.3 yd by 60.2 yd. Lot B measures 125.5 yd by 204.8 yd. What is the total area of the two lots?

15. _____

16. Kylie purchased 3 shirts for $16.59 each and paid with a $100 bill. How much change should she receive?

16. _____

17. A play area is 57 ft by 35 ft. It is covered with grass except for the area of the sand box which is 12.5 ft by 17.5 ft. How much area is covered with grass?

17. _____

94

18. In order to make money on loans, financial institutions are paid back more money than they loan. Gilbert borrows $115,000 to buy a house and agrees to make monthly payments of $671.11 for 30 yr. How much does he pay altogether? How much more does he pay than the amount of the loan?

18. _____

19. Cora's cell phone plan is $40 per month for 400 anytime minutes. Minutes in excess of four hundred are charged at the rate of $0.175 per minute. One month Cora used her phone for 668 minutes. How much was she charged?

19. _____

20. Katie paid $2.89 for a bag of 12 candy bars. Find the cost per candy bar. Round to the nearest tenth of a cent.

20. _____

21. Ben is paid $17.76 per hour for the first 40 hr of work each week and $26.63 per hour for hours in excess of 40 hr per week. One week he works 45 hr. How much is his pay for that week?

21. _____

22. The relay team had times of 20.6 seconds for the first leg, 22.4 seconds for the second leg, 22.2 seconds for the third leg, and 21.5 seconds for the fourth leg. Find the average time per leg.

22. _____

23. Brendan got 22 hits in 60 at-bats. What part of his at-bats were hits? Use decimal notation rounded to the nearest thousandth.

23. _____

24. What is the cost, in dollars, of 13.6 gal of gasoline at $3.089 per gallon? Round the answer to the nearest cent.

24. _____

25. A person who weighs 150 lb burns 13 calories per minute while cross-country skiing. In order to lose 1 lb, a person must burn about 3500 calories. For how long must this person cross-country ski in order to burn 3500 calories? Round to the nearest minute.

25. _____

Chapter 5 RATIO AND PROPORTION

5.1 Introduction to Ratios

Learning Objectives
Learning Objectives a Find fraction notation for ratios. b Simplify ratios.

Objective a   Find fraction notation for ratios.

Find fraction notation for the ratio. You need not simplify.

1. 3 to 4

1. _____

2. 7 to 9

2. _____

3. 149 to 150

3. _____

4. 85.1 to 105.2

4. _____

5. $5\frac{3}{4}$ to $6\frac{5}{8}$

5. _____

6. A rectangular carpet measures 12 ft by 10 ft. What is the ratio of length to width? of width to length?

6. _____

7. A sweater that originally sold for $35.95 was marked down to a sale price of $29.99. What is the ratio of the original price to the sale price? of the sale price to the original price?

7. _____

8. Monica has 6 grandsons and 5 granddaughters. What is the ratio of granddaughters to grandsons? of grandsons to granddaughters?

8. _____

Objective b   Simplify ratios.

Find the ratio of the first number to the second and simplify.

9. 12 to 15

9. _____

10. 8 to 10

10. _____

11. 24 to 36

11. _____

12. 16 to 18

12. _____

13. 96 to 100

13. _____

14. 5.6 to 6.4

14. _____

15. 3.6 to 5.4

15. _____

16. In this rectangle, find the ratio of length to width and width to length.

16. _____

0.3 cm

0.5 cm

98

Chapter 5 RATIO AND PROPORTION

5.2 Rates and Unit Prices

Learning Objectives
a   Give the ratio of two different measures as a rate.
b   Find unit prices and use them to compare purchases.

Key Terms
Use the vocabulary terms listed below to complete each statement in Exercises 1–2.

    rate   **unit price**

1. A _____ is a ratio used to compare two different kinds of measures.

2. A _____ is the ratio of price to the number of units.

Objective a   Give the ratio of two different measures as a rate.

Find the rate, or speed, as a ratio of distance to time in Exercises 3-4. Round to the nearest hundredth where appropriate.

3.   150 mi, 3 hr   3. _____

4.   834.3 m, 5.4 min   4. _____

5.   Craig's van travels 495.9 miles on 28.5 gallons of   5. _____
     gasoline. What is the rate in miles per gallon?

6.   Corey scored 276 points in 15 games. What was the rate in   6. _____
     points per game?

7.   Massachusetts has a population of 6,398,743 and is 10,555   7. _____
     sq mi in area. Find the population density (the rate per
     square mile). Round to the nearest one.

8.   To paint his house, Jerome needs 2 gal of paint for   8. _____
     700 sq ft. What is the rate in square foot per gallon?

Objective b   Find unit prices and use them to compare purchases.

Find the unit price of each brand in Exercises 9–12. Then, in each exercise, determine which brand is the better buy based on unit price.

9.

Grape Jelly			
Brand	**Size**	**Price**	**Unit Price**
A	24 oz	$1.79	
B	32 oz	$2.39	

9. _____

10.

Bottled Green Tea			
Brand	**Size**	**Price**	**Unit Price**
A	12 oz	$0.69	
B	64 oz	$3.09	

10. _____

11.

Coffee			
Brand	**Size**	**Price**	**Unit Price**
A	8 oz	$2.99	
B	11.5 oz	$4.09	
C	13 oz	$4.59	

11. _____

12.

Mixed Nuts			
Brand	**Size**	**Price**	**Unit Price**
A	9 oz	$4.25	
B	10.5 oz	$4.99	
C	13 oz	$6.19	

12. _____

100

Chapter 5 RATIO AND PROPORTION

5.3 Proportions

Learning Objectives
a Determine whether two pairs of numbers are proportional.
b Solve proportions.

Key Terms

Use the vocabulary terms listed below to complete each statement in Exercises 1–3.

    ratio   **cross products**

    proportional   **proportion**

1. Two numbers are _____ if they have the same _____.

2. A _____ is an equation which states that two ratios are equal.

3. In the equation $\dfrac{5}{8} = \dfrac{10}{16}$, we call $5 \cdot 16$ and $8 \cdot 10$ _____ .

Objective a   Determine whether two pairs of numbers are proportional.

Determine whether the two pairs of numbers are proportional.

4. 8, 12 and 12, 18

4. _____

5. 16, 5 and 13, 2

5. _____

6. 15, 7 and 45, 21

6. _____

7. $4\dfrac{1}{2}$, 8 and $11\dfrac{1}{4}$, 20

7. _____

8. 1.2, 2 and 6, 6.8

8. _____

Objective b   Solve proportions.

Solve.

9. $\dfrac{3}{8} = \dfrac{x}{5}$

9. _____

10. $\dfrac{18}{10} = \dfrac{45}{x}$

10. _____

11. $\dfrac{y}{0.6} = \dfrac{0.12}{0.4}$

11. _____

12. $\dfrac{3\frac{1}{2}}{m} = \dfrac{\frac{1}{2}}{\frac{1}{8}}$

12. _____

13. $\dfrac{10.4}{6.5} = \dfrac{8.2}{t}$

13. _____

14. $\dfrac{x}{13} = \dfrac{5.1}{2}$

14. _____

15. $\dfrac{x}{19} = \dfrac{3}{7}$

15. _____

102

Chapter 5 RATIO AND PROPORTION

5.4 Applications of Proportions

Learning Objectives
a   Solve applied problems involving proportions.

Objective a   Solve applied problems involving proportions.

Solve.

1. A 12-oz can of cola contains 140 calories. How many   1. _____
    calories are there in an 8-oz glass?

2. Tula drove her car 8000 mi in the first 5 months. At this   2. _____
    rate how many miles would she drive in a year?

3. The ratio of adults and children was 3 to 8. If there were   3. _____
    32 children, how many adults were there?

4. On a map, $\frac{1}{2}$ in. represents 75 mi. If two cities are $1\frac{3}{4}$ in.   4. _____
    apart on the map, how far apart are they in reality?

5. To determine the number of trout in a pond, a homeowner   5. _____
    catches 20 trout, tags them and releases them. Later, 12
    trout are caught, and it is found that 5 of them are tagged.
    Estimate of the number of trout in the pond.

6. The ratio of students to teachers at a private college is 16 to 1. If 48 students register for Introduction to Psychology, how many sections of the course would you expect to see offered?

6. _____

7. A quality-control inspector examined 100 pens and found 4 of them to be defective. At this rate, how many defective pens will there be in a lot of 2700?

7. _____

8. A 9-lb turkey breast contains 39 servings of meat. How many pounds of turkey breast would be needed for 65 servings?

8. _____

9. A high school history teacher must grade 33 reports. He can grade 5 reports in 75 min. At this rate, how long will it take him to grade all 33 reports?

9. _____

10. Every 7 pages of an author's manuscript corresponds to 6 published pages. How many published pages will a 728-page manuscript become?

10. _____

104

Chapter 5 RATIO AND PROPORTION

5.5 Geometric Applications

Learning Objectives
a   Find lengths of sides of similar triangles using proportions.
b   Use proportions to find lengths in pairs of figures that differ only in size.

Key Terms
Use the vocabulary terms listed below to complete each statement in Exercises 1–2.

    ratio   **shape**

1.   Similar triangles have the same _____ .

2.   The lengths of corresponding sides of similar triangles have the same
   _____ .

Objective a   Find lengths of sides of similar triangles using proportions.

Find the missing lengths.
3.   If $\triangle FGH \sim \triangle QRS$, find RQ and RS.   3. _____

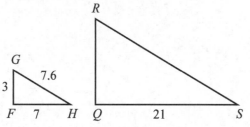

4.   If $\triangle XYZ \sim \triangle CDE$, find CD and DE.   4. _____

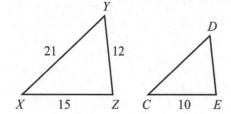

5. If $\overline{PQ} \parallel \overline{LM}$, find QN.

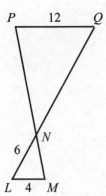

6. If $\overline{JK} \parallel \overline{BC}$, find KX.

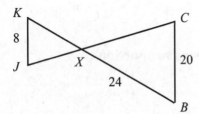

7. How high is a tree that casts a 50-ft shadow at the same time that a 6-ft man casts a 4-ft shadow?

8. How high is a flagpole that casts a 30-ft shadow at the same time a 30-ft tower casts a 20-ft shadow?

9. Find the distance d. Assume that the ratio of d to 13 m is   **9.** _____
the same as the ratio of 24 m to 8 m.

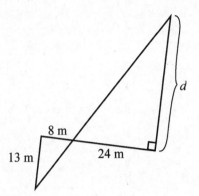

Objective b   Use proportions to find lengths of pairs of figures that differ only in size.

In each of Exercises 10–14, the sides in each pair of figures are proportional. Find the missing lengths.

10.   **10.** _____

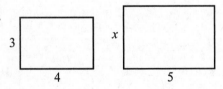

11.   **11.** _____

12.

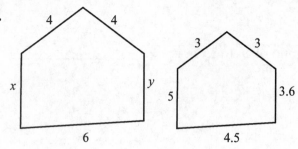

12. _____

13.

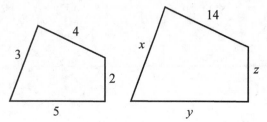

13. _____

14.

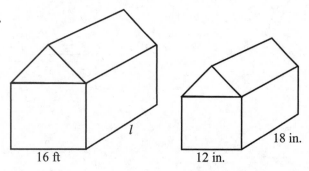

14. _____

Chapter 6 PERCENT NOTATION

6.1 Percent Notation

Learning Objectives
a   Write three kinds of notation for a percent.
b   Convert between percent notation and decimal notation.

Key Terms
Use the vocabulary terms listed below to complete each statement in Exercises 1–2.

     percent   **ratio**

1.   _____ means per hundred.

2. $n\%$ can be expressed as the _____ of n to 100.

Objective a   Write three kinds of notation for a percent.

Write three kinds of notation for the given percent ($n\% = \dfrac{n}{100}$; $n\% = n \times \dfrac{1}{100}$; $n\% = n \times 0.01$).

3.   45%   3. _____

4.   26.2%   4. _____

5.   11.1%   5. _____

6.   240%   6. _____

Objective b   Convert between percent notation and decimal notation.

Find decimal notation.

7.   23%   7. _____

8.   46.2%   8. _____

9.   31.76%   9. _____

10. 20%   10. _____

11. 6%   11. _____

12. 100%

13. 800%

14. 0.9%

15. 0.45%

16. 0.05%

17. 4.2%

18. $12\frac{1}{2}\%$

12. _____

13. _____

14. _____

15. _____

16. _____

17. _____

18. _____

Find decimal notation for the percent notation in the sentence.

19. Last year, 35% of the seeds Simon planted sprouted.

20. Collett earned a bonus of $8\frac{1}{2}\%$ of her sales.

19. _____

20. _____

Find percent notation.

21. 0.93

22. 0.04

23. 2.5

24. 0.7

25. 0.001

26. 0.025

27. 0.4876

28. 0.00068

21. _____

22. _____

23. _____

24. _____

25. _____

26. _____

27. _____

28. _____

Find percent notation for the decimal in the sentence.

29. About 0.75 of the Carsons' meals are vegetarian.

30. Recently, 0.04 of Kevin's classmates said they slept 12 hours the previous night.

29. _____

30. _____

110

Chapter 6 PERCENT NOTATION

6.2 Percent and Fraction Notation

Learning Objectives
a   Convert from fraction notation to percent notation.
b   Convert from percent notation to fraction notation.

Key Terms
Use the vocabulary terms listed below to complete each statement in Exercises 1–2.

decimal equivalent   **fraction equivalent**

1. The _____ to 0.4 is $\frac{2}{5}$.

2. The _____ to $\frac{1}{6}$ is $0.1\overline{6}$

Objective a   Convert from fraction notation to percent notation.

Find percent notation.

3. $\frac{17}{100}$

4. $\frac{8}{100}$

5. $\frac{3}{10}$

6. $\frac{1}{4}$

7. $\frac{5}{8}$

8. $\frac{5}{6}$

3. _____

4. _____

5. _____

6. _____

7. _____

8. _____

9. $\dfrac{15}{16}$

10. $\dfrac{7}{20}$

11. $\dfrac{19}{25}$

9. _____

10. _____

11. _____

Find percent notation for the fraction in the sentence.

12. John got $\dfrac{19}{20}$ of the problems correct.

12. _____

13. The Lamberts spent $\dfrac{3}{50}$ of their income on higher education courses.

13. _____

Write percent notation for the fractions in this pie chart.

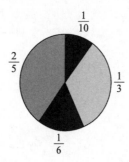

14. $\dfrac{1}{10}$

15. $\dfrac{1}{3}$

16. $\dfrac{2}{5}$

17. $\dfrac{1}{6}$

14. _____

15. _____

16. _____

17. _____

Objective b   Convert from percent notation to fraction notation.

Find fraction notation. Simplify.

18. 65%

19. 37.5%

20. 83.$\overline{3}$%

21. 0.4%

22. $11\dfrac{1}{9}\%$

23. 350%

24. 0.053%

18. _____

19. _____

20. _____

21. _____

22. _____

23. _____

24. _____

Find fraction notation for the percent notation in the following bar graph.

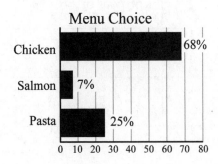

Menu Choice

25. 68%

26. 7%

27. 25%

25. _____

26. _____

27. _____

113

Find fraction notation for the percent notation in the sentence.

28. One 12-oz can of cola provides 13% of the total recommended daily allowance of carbohydrates.

28. _____

29. Quinn's electric bill was 3.5% lower in April than it was in March.

29. _____

Complete the table.

30.

Fraction Notation	Decimal Notation	Percent Notation
$\dfrac{1}{6}$		
	0.25	
		30%
$\dfrac{2}{5}$		
	0.375	
		$66.\overline{6}\%$, or $66\dfrac{2}{3}\%$
$\dfrac{7}{8}$		
	$0.8\overline{3}$	

Chapter 6 PERCENT NOTATION

6.3 Solving Percent Problems Using Percent Equations

Learning Objectives
a   Translate percent problems to percent equations.
b   Solve basic percent problems.

Key Terms
Use the vocabulary terms listed below to complete each statement in Exercises 1–2.

    amount   **base**

1.   The _____ in a percent problem refers to the number you are taking the percent of.

2.   The _____ in a percent problem refers to the result of taking a percent.

Objective a   Translate percent problems to percent equations.

Translate to an equation. Do not solve.

3.   What is 15% of 60?   3. _____

4.   14% of 20 is what?   4. _____

5.   25 is what percent of 125?   5. _____

6.   What percent of 24 is 5?   6. _____

7.   20 is 45% of what?   7. _____

8.   62.5% of what is 38?   8. _____

Objective b   Solve basic percent problems.

Translate to an equation and solve.

9.   What is 35% of 200?   9. _____

10. 125% of 42 is what?   10. _____

11. What is 4% of $800?

11. _____

12. 5.6% of 18 is what?

12. _____

13. $37\frac{1}{2}\%$ of 240 is what? (Hint $37\frac{1}{2}\% = \frac{3}{8}$.)

13. _____

14. $24 is what percent of $80?

14. _____

15. 800 is what percent of 250?

15. _____

16. What percent of $640 is $400?

16. _____

17. What percent of 75 is 100?

17. _____

18. 45 is 60% of what?

18. _____

19. 12% of what is $54?

19. _____

20. 26.88 is 48% of what?

20. _____

21. 80% of what is 6?

21. _____

22. What is $66\frac{2}{3}\%$ of 15?

22. _____

23. What is 6.5% of $856?

23. _____

24. 100% of what is $18.02?

24. _____

116

Chapter 6 PERCENT NOTATION

6.4 Solving Percent Problems Using Proportions

Learning Objectives
a   Translate percent problems to proportions.
b   Solve basic percent problems.

Key Terms

Use the vocabulary terms listed below to complete each statement in Exercises 1–2.

part   **whole**

1. When translating a percent problem to a proportion, the _____
    corresponds to the base.

2. When translating a percent problem to a proportion, the _____
    corresponds to the amount.

Objective a   Translate percent problems to proportions.

Translate to a proportion. Do not solve.

3. What is 28% of 36?   3._____

4. 55% of 40 is what?   4._____

5. 2.2 is what percent of 20?   5._____

6. What percent of 5.4 is 0.9?   6._____

7. 19 is 20% of what?   7._____

8. 150% of what is 32?   8._____

Objective b Solve basic percent problems.

Translate to a proportion and solve.

9. What is 35% of 60?

9. _____

10. 80% of 420 is what?

10. _____

11. What is 5% of 900?

11. _____

12. 2.4% of 12 is what?

12. _____

13. $66 is what percent of $150?

13. _____

14. 312 is what percent of 100?

14. _____

15. What percent of $864 is $216?

15. _____

16. What percent of 36 is 33?

16. _____

17. $27 is 75% of what?

17. _____

18. 70% of what is 126?

18. _____

19. 43.5 is 27% of what?

19. _____

20. 50% of what is 9?

20. _____

21. What is $83.\overline{3}$% of 300?

21. _____

22. What is 4.5% of $3924?

22. _____

Chapter 6 PERCENT NOTATION

6.5 Applications of Percent

Learning Objectives
a   Solve applied problems involving percent.
b   Solve applied problems involving percent of increase or decrease.

Key Terms
Use the vocabulary terms listed below to complete each statement in Exercises 1–2.

decrease   **percent of decrease**

If the population in a town decreases, then

1. $\dfrac{\text{Original Population} - \text{New Population}}{\text{Original Population}} =$ _____ in Population

2. Original Population – New Population = _____ in Population

Objective a   Solve applied problems involving percent.

Solve.

3. There are 45 students in third grade. These students constitute 36% of the school population. How many students are in the school?

3. _____

4. During the spring semester, 80 students took an introductory level fine arts class. The number enrolled in each class is shown in the table below. What percent of the students are enrolled in each class?

4. _____

Class	Number Enrolled
Intro to Modern Dance	12
Drawing I	26
Intro to Jazz	15
Chorale I	18
Clay	9

5. Pat bought a rare baseball card for $48. He then sold it for   5. _____
    150% of the price for which he bought it. For how much
    did he sell it?

6. There are 8280 people living in Calais. Of these, 60%   6. _____
    have lived there more than 20 yr. How many have lived in
    Calais more than 20 yr?

7. A gardener makes 750 mL of a solution of water and plant   7. _____
    food concentrate; 5% is plant food concentrate. How many
    milliliters are plant food concentrate? water?

8. Glenna planted 420 fir trees. A storm damaged 29% of   8. _____
    them. How many trees were damaged in the storm? Round
    to the nearest whole number.

9. On a 60 question test, Charles got 88% correct. How many   9. _____
    did he get correct? incorrect? (Partial credit was given on
    some items.)

10. On a test, Mael got 18.25 items, or 73% of the items   10. _____
    correct. How many items were on the test? (There was
    partial credit on some items.)

11. Of the 40 hr in her workweek, Jayne spends 25 of them   11. _____
    working directly with customers. What percent is this?

12. Cliff earns bonus dollars at his hardware store for 3% of   12. _____
    the value of purchases of non-consumable items. One day
    he purchases $124.76 worth of goods. Of this purchase,
    $42.52 is in consumable items. Find the number of bonus
    dollars Cliff earned. Round to the nearest cent.

120

Objective b   Solve applied problems involving percent of increase or decrease.

Solve.

13. The amount in a savings account increased from $500 to   13. _____
    $540. What was the percent of increase?

14. During a sale, a price of the book decreased from $32 to   14. _____
    $24. What was the percent of decrease?

15. The population of Maryville increased from 45,206 to   15. _____
    46,384. What was the percent of increase?

16. A person earning $18.70 per hour receives a 6% raise.   16. _____
    What is the new hourly rate?

17. A computer hardware item purchased for $800 depreciates   17. _____
    in value 30% each year after purchase. What is the
    depreciated value of the item 1 yr after purchase? 2 yr
    after purchase?

18. A lamp has a regular price of $84. It is on sale for $75.   18. _____
    What is the percent of decrease?

19. Betty adds a tip of 17% of the cost of her meal to her   19. _____
    credit card charge. What is the total amount charged if the
    cost of the meal, without tip, is $12? $50?

20. The population of Enwood decreased from 38,576 to 35,413 during the last decade.
    a)   What is the percent of decrease?

20. _____

    b)   If the trend continues and the same percent of decrease occurs during the next decade, what will be the population at the end of the next decade?

21. The population of Carson increased from 12,492 to 13,822.
    a)   Find the increase.

21. _____

    b)   Find the percent of increase.

22. A piece of office equipment depreciates 30% per year after purchase. One year after purchase, a piece of equipment is worth $2345. How much was it worth originally?

22. _____

Chapter 6 PERCENT NOTATION

6.6 Sales Tax, Commission, and Discount

Learning Objectives
a   Solve applied problems involving sales tax and percent.
b   Solve applied problems involving commission and percent.
c   Solve applied problems involving discount and percent.

Key Terms
Use the vocabulary terms listed below to complete each statement in Exercises 1–3.

commission	**discount**	**sales tax**
simple interest	**compound interest**	

1.   One earns a _____, when one is paid a percent of sales.

2.   A _____ is an amount subtracted from an original price.

3.   _____ is a percent of the purchase price of an item, which is paid to a government.

Objective a   Solve applied problems involving sales tax and percent.

Solve.

4.   The sales tax rate in Kentucky is 6%. How much tax would be charged on the purchase of a couch which costs $1299?

4. _____

5.   The sales tax rate in Arizona is 5.6%. How much tax is charged on the purchase of 6 CDs at $18 apiece? What is the total price?

5. _____

6.   The sales tax on the purchase of a rocking chair that sells for $459 is $27.54. What is the sales tax rate?

6. _____

123

7. The sales tax on the purchase of a new car is $832.50 and the sales tax rate is 4.5%. Find the purchase price.

7. _____

8. The sales tax rate for the municipality of Manchester is 1% and the state sales tax rate is 5.75%. Find the total amount paid for two computer desks at $259 each.

8. _____

Objective b   Solve applied problems involving commission and percent.

9. Barry's commission rate is 7%. What is the commission on the sale of $12,500 worth of merchandise?

9. _____

10. Keturah earns $57 selling $380 worth of books. What is her commission rate?

10. _____

11. An auction house's commission rate is 30%. They receive a commission of $2460. How many dollars worth of goods did they auction?

11. _____

12. Tess' commission is increased according to how much she sells. She receives a commission of 6% for the first $5000 and 12% on the amount over $5000. What is the total commission on sales of $12,000?

12. _____

Objective c   Solve applied problems involving discount and percent.

Complete the table by filling in the missing numbers.

	Marked Price	Rate of Discount	Discount	Sale Price
13.	$120	20%		
14.	$5,625	15%		
15.		30%	$33.60	
16.		18%	$76.50	
17.	$250		$62.50	
18.	$16,000		$1920	

124

Chapter 6 PERCENT NOTATION

6.7 Simple and Compound Interest; Credit Cards

Learning Objective
a   Solve applied problems involving simple interest.
b   Solve applied problems involving compound interest.
c   Solve applied problems involving interest rates on credit cards.

Key Terms
Use the vocabulary terms listed below to complete each statement in Exercises 1–3.

> APR   simple interest
> compound interest

1.   When interest is paid on interest, we call it _____.

2.   Principal×Rate×Time = _____.

3.   The _____ is the yearly interest rate associated with a credit card.

Objective a   Solve applied problems involving simple interest.

Find the simple interest.

	Principal	Rate of Interest	Time	Simple Interest
4.	$500	3%	1 year	
5.	$5,400	9.5%	$\frac{1}{2}$ year	
6.	$120,000	$5\frac{7}{8}$ %	$\frac{1}{4}$ year	

Solve. Assume that simple interest is being calculated in each case.

7.   Kayla's Collectibles borrows $14,200 at 6% for 60 days.   7. _____
     Find (a) the amount of interest due and (b) the total
     amount to be paid after 60 days.

8. Hair Expressions borrows $6500 at 4.5% for 30 days. Find   **8.** _____
    (a) the amount of interest due and (b) the total amount to
    be paid after 30 days.

Objective b   Solve applied problems involving compound interest.

*Interest is compounded annually. Find the amount in the account after the given length
of time. Round to the nearest cent.*

	Principal	Rate of Interest	Time	Amount in the Account
9.	$750	4%	2 years	
10.	$5,000	7.2%	4 years	
11.	$80,000	$8\frac{1}{4}\%$	10 years	

*Interest is compounded semiannually. Find the amount in the account after the given
length of time. Round to the nearest cent.*

	Principal	Rate of Interest	Time	Amount in the Account
12.	$2000	8.9%	3 years	
13.	$150,000	$6\frac{1}{2}\%$	20 years	

Objective c   Solve applied problems involving interest rates on credit cards.

Solve.

14. At the end of her sophomore year, Jenna has a balance of   **14. a)**_____
$3256.78 on a credit card with an annual percentage rate
(APR) of 19.8%. She decides not to make additional
purchases with her card until she has paid off the balance.

  a) Many credit cards require a minimum monthly
     payment of 2% of the balance. What is Jenna's
     minimum payment on a balance of $3256.78? Round
     the answer to the nearest dollar.

  b) Find the amount of interest and the amount applied to   **b)**_____
     reduce the principal in the minimum payment found
     in part (a).

  c) If Jenna had transferred her balance to a card with an
     APR of 15.5%, how much of her first payment would   **c)**_____
     be interest and how much would be applied to reduce
     the principal?

  d) Compare the amounts for 15.5% from part (c) with   **d)**_____
     the amounts for 19.8% from part (b).

15. At the end of his sophomore year, Jason has a balance of $5004.63 on a credit card with an annual percentage rate (APR) of 21.5%. He decides not to make additional purchases with his card until he has paid off the balance.

a) Many credit cards require a minimum monthly payment of 2% of the balance. What is Jason's minimum payment on a balance of $5004.63? Round the answer to the nearest dollar.

15. a)_____

b) Find the amount of interest and the amount applied to reduce the principal in the minimum payment found in part (a).

b)_____

c) If Jason had transferred his balance to a card with an APR of 13.9%, how much of his first payment would be interest and how much would be applied to reduce the principal?

c)_____

d) Compare the amounts for 13.9% from part (c) with the amounts for 21.5% from part (b).

d)_____

128

Chapter 7 DATA, GRAPHS, AND STATISTICS

7.1 Averages, Medians, and Modes

Learning Objectives
a   Find the average of a set of numbers and solve applied problems involving averages.
b   Find the median of a set of numbers and solve applied problems involving medians.
c   Find the mode of a set of numbers and solve applied problems involving modes.
d   Compare two sets of data using their means.

Key Terms
Use the vocabulary terms listed below to complete each statement in Exercises 1–4.

   statistic   **mean**   **median**   **mode**

1.   To find the _____ of a set of data, we add the numbers and then divide
     by the number of data items.

2.   The mean is an example of a _____, a number describing a set of data.

3.   The _____ of a set of data is the number or numbers that occur most
     often.

4.   The _____ is the middle number or the average of the two middle
     numbers in a list that has been ordered from smallest to largest.

Objective a   Find the average of a set of numbers and solve applied problems involving averages.

For each set of numbers, find the average, or mean.

5.   7, 7, 7, 8, 8, 8, 25   5. _____

6.   6, 8, 12, 15, 21   6. _____

7.   3.8, 1.2, 7.6, 5.6, 1.2, 7.6   7. _____

8. $25, $30, $25, $10   8. _____

9. Terry gets 188 mi of city driving on 8 gal of gasoline. What is the average number of miles expected per gallon—that is, what is her vehicle's gas mileage?

9. _____

10. Find the grade point average for the student with the course load shown in the table. Assume the grade point values are 4.0 for an A, 3.0 for a B, and so on. Round to the nearest tenth.

10. _____

Grade	Number of Credit Hours in Course
A	3
B	4
C	3
B	3
A	1

11. To get an A in history, Brent must score an average of 90 on 5 papers. Scores on the first four papers were 82, 95, 88, and 88. What is the lowest score that Brent can get on his last paper and still receive an A?

11. _____

Objective b   **Find the median of a set of numbers and solve applied problems involving medians.**

For each set of numbers, find the median.

12. 10, 10, 10, 13, 13, 15, 15

12. _____

13. 4, 12, 10, 16

13. _____

14. 2, 2, 8, 7, 3

14. _____

15. 2.5, 1.9, 4.2, 3.2, 3.8, 2.9

15. _____

16. $225, $250, $175, $250, $225

16. _____

130

17. The following table shows the mean temperature in July in New York City for several years. Find the median of these temperatures.

17. _____

Year	Mean July Temperature (in degrees Fahrenheit)
1965	67.7
1975	69.1
1985	69.5
1995	73.6
2005	75.1

18. The following prices per gallon for milk were found at five supermarkets:

$3.29 $3.35 $3.09 $3.35 $3.25.

Find the median price per gallon.

18. _____

Objective c   Find the mode of a set of numbers and solve applied problems involving modes.

For each set of numbers find any modes that exist.

19. 5, 5, 5, 1, 2, 1, 2

19. _____

20. 3.5, 2.9, 3.2, 3.4

20. _____

21. $30, $25, $25, $20, $30

21. _____

22. 5, 8, 10, 6, 7, 8

22. _____

23. 0, 2, 1, 3, 0, 1, 0, 3

23. _____

131

24. The following bar graph shows the annual precipitation, in inches, in Phoenix, Arizona, for several recent years. Find any modes that exist.

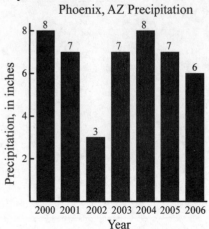

Phoenix, AZ Precipitation

Objective d   Compare two sets of data using their means.

Solve.

25. An experiment is performed to compare the lives of two types of batteries. Several batteries of each type were tested and the results are listed in the following table. On the basis of this test, which battery is better?

Battery A: lifetime (in hours)			Battery B: lifetime (in hours)		
18	20	18	17	20	20
22	19	21	21	19	21
20	23	19	16	19	20
18	17	18	20	18	19

26. An experiment is performed to compare the tread life of two types of sneakers. Several pairs of each type were tested and the results are listed in the following table. On the basis of this test, which sneaker has the better tread life?

26. _____

Sneaker A: tread life (in miles)			Sneaker B: tread life (in miles)		
360	400	420	410	420	400
420	430	450	390	430	420
390	410	440	400	420	410

27. An experiment is conducted to determine which of two ice cream flavors tastes better. Taste testers ate each ice cream flavor and gave it a rating from 1 to 10, where 10 is the best. The results are in the following table. On the basis of this test, which flavor tastes better?

27. _____

Flavor A: Pineapple/Mango			Flavor B: Strawberry/Mango		
7	8	6	9	8	7
9	7	7	10	9	8
8	10	8	6	7	8
6	7	10	8	6	5

28. An experiment is conducted to determine which of two coffee flavors tastes better. Taste testers drank each coffee and gave it a rating from 1 to 10, where 10 is the best. The results are given in the following table. On the basis of this test, which coffee tastes better?

28. _____

Coffee A: Hint of Mint			Coffee B: Maple Delight		
6	8	7	6	6	7
5	6	10	8	5	9
4	7	8	6	7	7
6	5	9	8	7	6

Chapter 7 DATA, GRAPHS, AND STATISTICS

7.2 Tables and Pictographs

Learning Objectives
a   Extract and interpret data from tables.
b   Extract and interpret data from pictographs.

Key Terms

Use the vocabulary terms listed below to complete each statement in Exercises 1–2.

pictograph   **key**

1. A _____ is a graph that uses symbols to represent the amounts.

2. The _____ on a _____ tells what each symbol represents.

Objective a   Extract and interpret data from tables.

Use the following table, which lists information about several countries, for Exercises 3–7.

Country	Form of Government	Area (in sq km)	Population	Median Age (in yr)	Literacy Rate
Argentina	Republic	2,766,890	40,301,297	29.9	97.1%
Chad	Republic	1,284,000	9,885,661	16.3	47.5%
Denmark	Constitutional Monarchy	43,094	5,468,120	40.1	99%
Georgia	Republic	69,700	4,646,003	38.0	100%
St. Lucia	Parliamentary Democracy	616	170,649	25.6	90.1%
Vietnam	Communist State	329,560	85,262,356	26.4	90.3%

3. Find the median age of a person in Denmark.   3. _____

4. Which countries listed are republics?   4. _____

5. Which countries listed have a literacy rate above 95%?   5. _____

6. How much greater is the population of Chad than Denmark?   6. _____

7. What are the average, the median, and the mode of the literacy rates?   7. _____

Use the following table, which lists information about several U.S. states, for Exercises 8–15.

State	Year Entered the Union	Area (in sq mi)	Population Density (per sq mi)	Median Income (2004–2005)
Alabama	1819	52,419	88.7	$46,071
California	1848	163,696	227.5	$51,312
Delaware	1776	2489	418.5	$50,445
Idaho	1890	83,570	16.5	$45,009
Michigan	1858	86,939	63.6	$44,801
Ohio	1803	40,948	279.3	$44,349

The population of a state is its area, in sq mi, times its population density, in persons per sq mi. Find the population of:

8. Delaware.

8. _____

9. Michigan.

9. _____

10. What state has a median income of $44,349?

10. _____

11. What year did Idaho enter the Union?

11. _____

12. How much greater is the population density per sq mi of California than of Alabama?

12. _____

13. Which state listed has the least area?

13. _____

14. Which two states listed have the largest difference in population density? What is the difference?

14. _____

15. Find the median income of the state with the largest area.

15. _____

136

Objective b   Extract and interpret data from pictographs.

The following pictograph shows the highest recorded temperature in several states. Use the pictograph for Exercises 16–20.

State	Highest Recorded Temperature
Alaska	☼ ☼ ☼ ☼ ☼
Hawaii	☼ ☼ ☼ ☼ ☼
Maine	☼ ☼ ☼ ☼ ☼ (
Nevada	☼ ☼ ☼ ☼ ☼ ☼ (
North Carolina	☼ ☼ ☼ ☼ ☼ (
South Dakota	☼ ☼ ☼ ☼ ☼ ☼
☼ = 20°F	

16. What is the highest recorded temperature in North Carolina?

16. _____

17. Of the states shown, which has the highest recorded temperature?

17. _____

18. How much higher is the highest recorded temperature in South Dakota than in Maine?

18. _____

19. Which two states have approximately equal highest recorded temperatures?

19. _____

20. Which state's highest recorded temperature is about 120°F?

20. _____

The following pictograph shows shampoo sales for a company for several years. Use the pictograph for Exercises 21–26.

Year	Shampoo Sales
2002	🧴🧴🧴🧴🧴🧴
2003	🧴🧴🧴🧴🧴
2004	🧴🧴🧴🧴🧴
2005	🧴🧴🧴🧴🧴🧴🧴
2006	🧴🧴🧴🧴🧴🧴🧴🧴

🧴 = 1000 bottles sold

21. What were the shampoo sales in 2005?

21. _____

22. In which year were approximately 5000 bottles of shampoo sold?

22. _____

23. In which year were shampoo sales the least?

23. _____

24. How many more bottles of shampoo were sold in 2006 than in 2002? What was the percent of increase?

24. _____

25. Between which two years was there a decline in sales of shampoo?

25. _____

26. Between which two years was the increase in sales of shampoo approximately 500 bottles?

26. _____

Chapter 7 DATA, GRAPHS, AND STATISTICS

7.3 Bar Graphs and Line Graphs

Learning Objectives
a   Extract and interpret data from bar graphs.
b   Draw bar graphs.
c   Extract and interpret data from line graphs.
d   Draw line graphs.

Key Terms
Use the vocabulary terms listed below to complete each statement in Exercises 1–2.

   vertical   **horizontal**

1.   A _____ axis goes from left to right.

2.   A _____ axis goes up and down.

Objective a   Extract and interpret data from bar graphs.

The following horizontal bar graph shows the maximum number of consecutive hours allowed for persons in various occupations to work. Use the bar graph for Exercises 3–5.

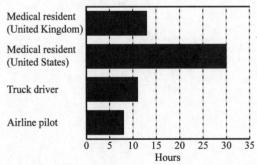

Maximum Consecutive Hours Allowed

3.   Which occupation has the fewest number of hours of   3. _____
     consecutive work allowed?

4.   How many consecutive hours of work are allowed for a   4. _____
     truck driver?

5.   How many more consecutive hours may a medical   5. _____
     resident in the United States work than a medical
     resident in the United Kingdom?

The following side-by-side vertical bar graph shows the life expectancy at birth for men and women in several countries. Use it for Exercises 6–9.

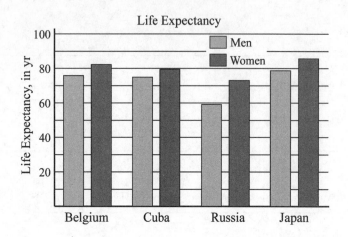

6. About how many years is a man in Cuba expected to live?

6. _____

7. About how many more years is a Belgian woman expected to live than a Belgian man?

7. _____

8. Which country had the greatest difference in life expectancy between men and women? About how many years was the difference?

8. _____

9. About how many more years was a man in Japan expected to live than a woman in Russia?

9. _____

Objective b   Draw bar graphs.

10. The following table lists the average number of pounds of fruit of several types consumed per person per year. Make a vertical bar graph of the data.

    10. _____

Fruit	Consumption in lbs
Apples	17
Bananas	25
Grapes	9
Oranges	11
Strawberries	6

Use the data and the bar graph in Exercise 10 to do Exercises 11–13.

11. Which type of fruit is the most popular?

    11. _____

12. What is the median of the number of pounds consumed of the 5 types of fruit listed?

    12. _____

13. What is the average of the number of pounds consumed of the 5 types of fruit listed?

    13. _____

14. The following table lists the average size of a new single-family home built in the U.S. for several years. Make a horizontal bar graph of the data.

14. _____

Year	Average Size, in sq ft.
1975	1645
1985	1785
1995	2095
2005	2434

Use the data and the bar graph in Exercise 14 to do Exercises 15–16.

15. How much larger was the average single-family home size in 2005 than in 1995?

15. _____

16. What was the percent of increase in the average single-family home size from 1975 to 2005?

16. _____

Objective c   Extract and interpret data from line graphs.

The following line graph shows average SAT verbal scores for several years. Use the line graph for Exercises 17–19.

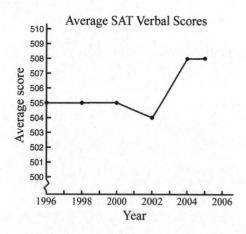

17. How much higher was the average SAT verbal score in 2005 than in 1998?

17. _____

18. What was the percent of increase in the average SAT verbal score from 2002 to 2004?

18. _____

19. In what year(s) was the average SAT verbal score 508?

19. _____

142

Objective d   Draw line graphs.

20. Make a line graph of the data in the following table, using   **20.** _____
the horizontal axis to scale "Year."

Year	Average Retail Price Per Gallon of Gasoline
1998	$1.07
2000	$1.52
2002	$1.39
2004	$1.90
2006	$2.62

Use the data and the line graph in Exercise 20 to do Exercises 21–23.

21. What was the median of the average retail prices per   **21.** _____
gallon of gasoline for the years given?

22. Between which two years was the greatest increase in   **22.** _____
average price per gallon of gasoline?

23. What was the percent of decrease in average price per   **23.** _____
gallon from 2000 to 2002?

24. Make a line graph of the data in the following table, using   **24.** _____
the horizontal axis to scale "Year."

Year	Median weekly earnings for full-time workers, in dollars
2002	608
2003	620
2004	638
2005	651
2006	671

Use the data and the line graph in Exercise 24 to do Exercises 25–27.

25. What was the average of the median weekly earnings for   **25.** _____
full-time workers for the years shown?

26. What was the percent of increase in weekly earnings from   **26.** _____
2005 to 2006?

27. How much more were the median weekly earnings for   **27.** _____
full-time workers in 2006 than in 2002?

Chapter 7 DATA, GRAPHS, AND STATISTICS

7.4 Circle Graphs

Learning Objectives
a   Extract and interpret data from circle graphs.
b   Draw circle graphs.

Key Terms
Use the vocabulary terms listed below to complete each statement in Exercises 1–2.

   pie chart   **100%**   **wedge**

1.   Another name for a circle graph is a _____.

2.   The total of all the sections, or _____ s, of a circle graph should be

     _____.

Objective a   Extract and interpret data from circle graphs.

The following circle graph shows the distribution of blood types in the United States.
Use the graph for Exercises 3–8.

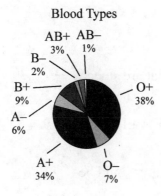

Blood Types

3.   What percent of people in the U.S. have blood type B+?   3. _____

4.   Together what percent of people in the U.S. have O+ or O-   4. _____
type blood?

5. If 500 people in a city donated a pint of blood at a blood drive, how many pints of A- blood would be expected to be collected?

5. _____

6. If 120 people in a state need blood one day, how many would be expected to need B+ blood type?

6. _____

7. What percent of all people in the U.S. do not have A+ blood type?

7. _____

8. What percent of all people in the U.S. do not have a positive blood type (that is, do not have O+, A+, B+, or AB+)?

8. _____

The following circle graph shows the distribution of types of material used for the exterior walls of newly built single-family homes. Use the graph for Exercises 9–14.

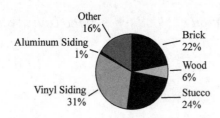

9. What percent of homes have exterior walls that are stucco?

9. _____

10. Together what percent of homes have exterior walls that are vinyl or aluminum siding?

10. _____

11. If 20,000 homes are built in a month, how many would be expected to have wood exterior walls?

11. _____

12. If 46,000 homes are built in a year, how many are expected to have brick exterior walls?

12. _____

13. What percent of all new single-family homes are built with exterior walls that are not vinyl siding?

13. _____

14. What percent of all new single-family homes are built with exterior walls that are not brick or stucco?

14. _____

146

Objective b   Draw circle graphs.

Use the given information to complete a circle graph. Note that each circle is divided into 100 equal sections.

15. The table below lists how often people claim they exercise.

15. <u>see graph</u>

Frequency	Percent
Daily	11.9%
More than 3 times per week, but less than daily	20.6%
2–3 times per week	29.3%
Once per week	19.2%
Never	16.2%
Other	2.8%

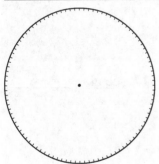

16. A survey was conducted and the table below lists respondents' favorite colors.

16. see graph

Color	Percent
Blue	42%
Purple	14%
Green	14%
Red	8%
Black	7%
Other	15%

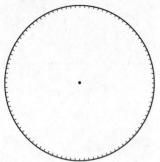

17. The table below lists the age of patients admitted to a hospital on a given weekend.

17. see graph

Age (in years)	Percent
0–4	8.8%
5–14	13.5%
15–40	15.6%
41–65	26.3%
Over 65	35.8%

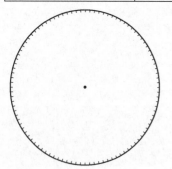

18. The table below lists the favorite desserts of the fifth-grade class at Johnson Academy.

    18. <u>see graph</u>

Dessert	Percent
Ice cream	26%
Cake	8%
Cookies	28%
Pie	17%
Fruit	15%
Other	6%

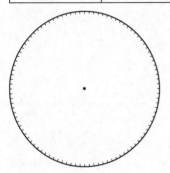

19. The table below lists the number of nights that guests stayed at a vacation lodge during a recent month.

    19. <u>see graph</u>

Length of Stay (in nights)	Percent
1	20%
2	30%
3–5	15%
5–7	25%
more than 7	10%

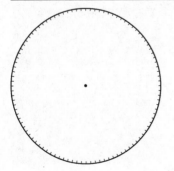

20. The table below lists the eye color of participants in a recent study.

20. see graph

Eye Color	Percent
Blue	21%
Brown	43%
Hazel	29%
Green	1%
Gray	3%
Amber	3%

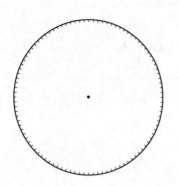

Chapter 8 MEASUREMENT

8.1 Linear Measures: American Units

Learning Objectives
a   Convert from one American unit of length to another.

Key Terms

Use the numbers listed below to complete each statement in Exercises 1–4.

       3   **12**   **36**   **5280**

1.   1 yard = _____ feet

2.   1 yard = _____ inches

3.   1 foot = _____ inches

4.   1 mile = _____ feet

Objective a   **Convert from one American unit of length to another.**

Complete.

5.   3 mi = ____ ft   **5.** _____

6.   36 ft = ____ yd   **6.** _____

7.   144 in. = ____ ft   **7.** _____

8. 5 yd = _____ in.

8. _____

9. 7.5 ft = _____ in.

9. _____

10. 4 mi = _____ yd

10. _____

11. 18 yd = _____ ft

11. _____

12. 3 mi = _____ in.

12. _____

13. 162 in. = _____ yd

13. _____

14. 2640 ft = _____ mi

14. _____

Chapter 8 MEASUREMENT

8.2 Linear Measures: The Metric System

Learning Objectives
a   Convert from one metric unit of length to another.

Key Terms
Use the numbers listed below to complete each statement in Exercises 1–3. Some of the numbers will not be used.

1	**10**	**100**	**1000**	**10,000**

1.   1 m = _____ cm

2.   1000 m = _____ km

3.   1 m = _____ mm

Objective a   Convert from one metric unit of length to another.

Complete. Do as much as possible mentally.

4.   6.2 km = _____ m   4. _____

5.   4000 cm = _____ m   5. _____

6.   2.5 m = _____ mm   6. _____

7.   14 m = _____ cm   7. _____

8.   3 km = _____ cm   8. _____

9. 150 mm = _____ m

9. _____

10. 5670 cm = _____ km

10. _____

11. 27 cm = _____ mm

11. _____

12. 3 mm = _____ cm

12. _____

13. 1350 m = _____ km

13. _____

14. 28,000 mm = _____ dam

14. _____

15. 43 dam = _____ dm

15. _____

16. 0.97 cm = _____ mm

16. _____

154

Chapter 8 MEASUREMENT

8.3 Converting Between American Units and Metric Units

Learning Objectives
a   Convert between American and metric units of length.

Objective a   Convert between American and metric units of length.

Complete. Answer may vary slightly depending on the conversion used.

1.   5 in. = ____ cm   1. _____

2.   4 km = ____ mi   2. _____

3.   3 yd = ____ m   3. _____

4.   6 m = ____ ft   4. _____

5.   2 m = ____ in.   5. _____

6.   7 ft = ____ m   6. _____

7. 180 cm = _____ in.

8. 1.5 yd = _____ m

9. 80 km/hr = _____mph.

Chapter 8 MEASUREMENT

8.4 Weight and Mass; Medical Applications

Learning Objectives
a   Convert from one American unit of weight to another.
b   Convert from one metric unit of mass to another.
c   Make conversions and solve applied problems concerning medical dosages.

Key Terms
Use the numbers listed below to complete each statement in Exercises 1–4. Some of the numbers will not be used.

16	1000	2000	$\dfrac{1}{1,000,000}$	1,000,000

1.   1 metric ton = _____ kilograms

2.   1 (American) ton = _____ pounds

3.   1 pound = _____ ounces

4.   1 g = _____ mcg

Objective a   Convert from one American unit of weight to another.

Complete.

5.   4 T = ____ lb   5. _____

6.   64 oz = ____ lb   6. _____

7. 56 oz = _____ lb

7. _____

8. 3000 lb = _____ T

8. _____

9. 2.5 lb = _____ oz

9. _____

10. 4.5 lb = _____ oz

10. _____

11. 5000 lb = _____ T

11. _____

12. 3.2T = _____ lb

12. _____

Objective b   Convert from one metric unit of mass to another.

Complete.

13. 1200 g = _____ kg

13. _____

14. 14 kg = _____ g

14. _____

158

15. 2 kg = _____ cg   **15.** _____

16. 4500 cg = _____ g   **16.** _____

17. 240 mg = _____ cg   **17.** _____

18. 3 t = _____ kg   **18.** _____

19. 6000 kg = _____ t   **19.** _____

20. 14 cg = _____ mg   **20.** _____

21. 3.6 kg = _____ g   **21.** _____

22. 1345 g = _____ kg   **22.** _____

Objective c   Make conversions and solve applied problems concerning medical dosages.

Complete.

23. 3 mg = _____ mcg   **23.** _____

24. 2.6 mg = _____ mcg   **24.** _____

25. 50 mcg = _____ mg   **25.** _____

26. 475 mcg = _____ mg   **26.** _____

Solve each of the following. (None of these medications should be taken without consulting your own physician)

27. Dr. Rodriguez prescribes a 0.4 mg dose of Flomax for a patient. How many micrograms are in this dose?

27.

28. Dr. Lee prescribes a 180 mg dose of Cardizem for a patient. How many micrograms are in this dose?

28.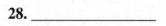

29. Dr. Ferraro prescribes an 18 mcg dose of Spiriva for a patient. How many milligrams are in this dose?

29.

30. Dr. Schwartz prescribes a 200 mcg dose of albuterol for a patient. How many milligrams are in this dose?

30. _____

Chapter 8 MEASUREMENT

8.5 Capacity; Medical Applications

Learning Objective
a   Convert from one unit of capacity to another.
b   Solve applied problems concerning medical dosages.

Key Terms
Use the numbers listed below to complete each statement in Exercises 1–4. Some of the numbers will not be used.

1   **2**   **4**   **8**   **16**

1.   4 qt = _____ gal

2.   1 qt = _____ pt

3.   1 pt = _____ fl oz

4.   1 cup = _____ fl oz

Objective a   Convert from one unit of capacity to another.

Complete.

5.   64 oz = ____ pt   **5.** _____

6.   32 oz = ____ cups   **6.** _____

7.   5 qt = ____ cups   **7.** _____

8.   16 pt = ____ gal   **8.** _____

9.   4 qt = ____ oz   **9.** _____

10. 38 L = ____ mL   **10.** _____

11. 514 mL = ____ L   **11.** _____

12. $120 \text{ cm}^3 = \underline{\hspace{1cm}} \text{ L}$

12. _____

13. $7 \text{ L} = \underline{\hspace{1cm}} \text{ cm}^3$

13. _____

Objective b Solve applied problems concerning medical dosages.

14. Charles received 60 mL per hour of normal saline solution. How many liters did he receive in a 48-hr period?

14. _____

15. Norma takes 48 mg of Tricor each day to battle high cholesterol. How many grams of this medication does she take in 30 days?

15. _____

16. Jenna takes 35 mg of Actonel once each week to battle osteoporosis. How many grams of this medication does she take in 12 weeks?

16. _____

17. Ricardo received 1800 mL of normal saline solution over a 24-hour period. How many liters is this?

17. _____

18. Dr. James wants a patient to receive 2.4 L of a normal glucose solution in a 24-hr period. How many milliliters per hour should the patient receive?

18. _____

19. Dr. Robbins wants a patient to receive 1.5 L of a normal glucose solution in a 12-hr period. How many milliliters per hour should the patient receive?

19. _____

Chapter 8 MEASUREMENT

8.6 Time and Temperature

Learning Objectives
a   Convert from one unit of time to another.
b   Convert between Celsius and Fahrenheit temperatures using the formulas.

Key Terms

Use the numbers listed below to complete each statement in Exercises 1–2. Some of the numbers will not be used.

0	32	100	212

1. $100°C =$ _____ $°F$

2. $32°F =$ _____ $°C$

Objective a   Convert from one unit of time to another.

Complete.

3.   240 sec = ____ min   3. _____

4.   5 wk = ____ days   4. _____

5.   7.5 days = ____ hr   5. _____

6.   1 wk = ____ hr   6. _____

7.   33,840 sec = ____ hr   7. _____

8.   372 hr = ____ days   8. _____

9.   2.5 wk = ____ min   9. _____

10. 4 yr = ____ days   10. _____

11. 504 hr = ____ wk

11. _____

12. 553 days = ____ wk

12. _____

13. What length of time is 10,080 min? Is it 1 hr, 1 day, 1 wk or 1 yr?

13. _____

Objective b   Convert between Celsius and Fahrenheit temperatures using the formulas.

Convert to Fahrenheit. Use the formula $F = \dfrac{9}{5} \cdot C + 32$ *or* $F = 1.8 \cdot C + 32$.

14. 15°C

14. _____

15. 50°C

15. _____

16. 80°C

16. _____

17. 45°C

17. _____

Convert to Celsius. Use the formula $C = \dfrac{5}{9}(F - 32)$ *or* $C = \dfrac{F - 32}{1.8}$.

18. 41°F

18. _____

19. 68°F

19. _____

20. 104°F

20. _____

21. 95°F

21. _____

164

Chapter 8 MEASUREMENT

8.7 Converting Units of Area

Learning Objectives
a   Convert from one American unit of area to another.
b   Convert from one metric unit of area to another.

Key Terms
Use the numbers listed below to complete each statement in Exercises 1–2. Some of the numbers will not be used.

3   6   9   12   24   144

1.   1 square foot = _____ square inches

2.   1 square yard = _____ square feet

Objective a   Convert from one American unit of area to another.

Complete.

3.   3 yd² = _____ ft²   3. _____

4.   8 ft² = _____ in²   4. _____

5.   3 acres = _____ ft²   5. _____

6.   2 mi² = _____ acres   6. _____

7.   720 in² = _____ ft²   7. _____

8.   36 ft² = _____ yd²   8. _____

9. 6 ft² = _____ in²

10. 14 yd² = _____ ft²

11. 45 ft² = _____ yd²

12. 1296 in² = _____ ft²

9. _____

10. _____

11. _____

12. _____

Objective b   Convert from one metric unit of area to another.

Complete.

13. 4.57 cm² = _____ mm²

14. 125,000 m² = _____ km²

15. 1560 cm² = _____ m²

16. 14 m² = _____ cm²

17. 2 km² = _____ m²

18. 1400 mm² = _____ m²

19. 164,976 cm² = _____ km²

20. 3 km² = _____ cm²

13. _____

14. _____

15. _____

16. _____

17. _____

18. _____

19. _____

20. _____

Chapter 9 GEOMETRY

9.1 Perimeter

Learning Objectives
a   Find the perimeter of a polygon.
b   Solve applied problems involving perimeter.

Key Terms
Use the vocabulary terms listed below to complete each statement in Exercises 1–3.

polygon   **square**   **rectangle**

1. A geometric figure with three or more sides is called a _____.

2. A _____ is any 4-sided polygon with four $90°$ angles.

3. A _____ is a polygon which has four $90°$ angles and four sides of the same length.

Objective a   Find the perimeter of a polygon.

Find the perimeter of the polygon.

4.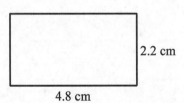

4. _____

5.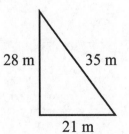

5. _____

6.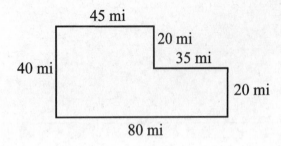

7.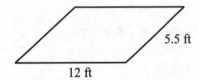

7. _____

8.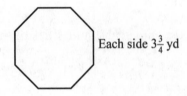

Each side $3\frac{3}{4}$ yd

8. _____

9.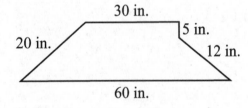

9. _____

Find the perimeter of the rectangle.

10. 6 ft by 9 ft

10. _____

11. $3\frac{1}{2}$ in. by $6\frac{3}{4}$ in.

11. _____

12. 5.76 cm by 8.42 cm

12. _____

Find the perimeter of the square.

13. 7 mi on a side

13. _____

14. 9.2 mm on a side

14. _____

15. $4\frac{1}{3}$ yd on a side

15. _____

Objective b   Solve applied problems involving perimeter.

Solve.

16. A homeowner decides to fence in a rectangular yard for his dog. The dimensions of the area to be fenced in are 20 ft by 35 ft. What is the perimeter of this area? If the fencing costs $2.19 per foot, how much will it cost him to fence in this area?

16. _____

17. A rectangular room is 12.5 ft by 14.5 ft. What is the perimeter of this room?

17. _____

18. A rectangular picture is 8.89 cm by 12.7 cm. What is the perimeter of the picture?

18. _____

19. A square quilt has sides of length 8.5 feet. What is the perimeter of the quilt?

19. _____

20. An L-shaped garden is to have a walkway built along its outer edge, as shown in the figure. Find the perimeter of the figure (including the walkway).

20. _____

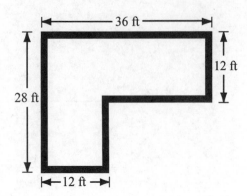

21. A landscaper plans to build a fence around a 16-m by 24-m play area.

   a) The posts of the fence are 2 m apart. How many posts will be needed?

21. a)_____

   b) The posts cost $3.20 each. What is the total cost for the posts?

b)_____

   c) The fence will surround all but 4 m of the play area, where there will be a two-panel gate. (A post is required between the panels, so the answers to parts (a) and (b) are not affected by the gate.) How long will the fence be?

c)_____

   d) The fencing costs $2.75 per meter. What will the cost of the fence be?

d)_____

   e) The two-panel gate costs $238. What is the total cost of the materials?

e)_____

170

Chapter 9 GEOMETRY

9.2 Area

Learning Objectives
a   Find the area of a rectangle and a square.
b   Find the area of a parallelogram, a triangle, and a trapezoid.
c   Solve applied problems involving areas of rectangles, squares, parallelograms, triangles, and trapezoids.

Key Terms

Use the formulas listed below to complete each statement in Exercises 1–3.

$$A = \frac{1}{2} \cdot b \cdot h \qquad A = b \cdot h \qquad A = \frac{1}{2} \cdot h \cdot (a+b)$$

1.   The formula for area of a parallelogram is _____.

2.   A formula for area of a trapezoid is _____.

3.   The formula for area of a triangle is _____.

Objective a   Find the area of a rectangle and a square.

Find the area.

4.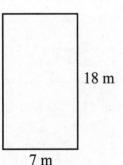
18 m
7 m

4. _____

5.
2.4 ft
2.4 ft

5. _____

6. 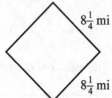 0.8 cm

3.2 cm

6. _____

7.

$8\frac{1}{4}$ mi

$8\frac{1}{4}$ mi

7. _____

Find the area of the rectangle.

8. 12 ft by 8 ft

8. _____

9. 6.7 m by 8.5 m

9. _____

10. $3\frac{3}{4}$ inches by $2\frac{5}{16}$ inches

10. _____

Find the area of the square.

11. 13 yd on a side

11. _____

12. 9.5 mm on a side

12. _____

13. $\frac{3}{4}$ mi on a side

13. _____

172

Objective b   Find the area of a parallelogram, a triangle, and a trapezoid.

Find the area.

14.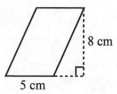

14. _____

15.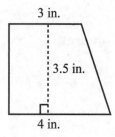

15. _____

16.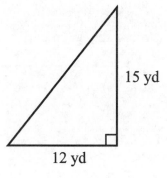

16. _____

17.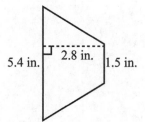

17. _____

18.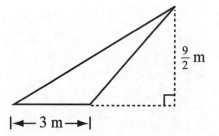

18. _____

173

19.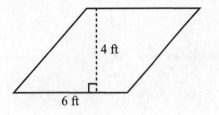

19. _____

20.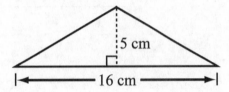

20. _____

Objective c   Solve applied problems involving areas of rectangles, squares, parallelograms, triangles, and trapezoids.

Solve.

21. A lot is 75 m by 50 m. A house 24 m by 10 m is built on the lot. How much area is left over for a lawn?

21. _____

22. A square garden $12\frac{1}{2}$ ft on a side is placed in a lawn that is 48 ft by 32 ft.

22. _____

   a)   Find the area of the lawn, excluding the garden.

   b)   It costs $0.35 per square foot to have a garden rototilled. Find the cost of rototilling the garden.

23. A city wants to put a sidewalk that is 2-ft wide around a community garden, as shown in the figure. The garden itself is 65 ft by 32 ft. Find the area of the sidewalk.

23. _____

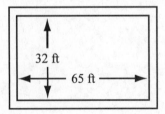

174

24. The page of a book is $10\frac{7}{8}$ in. by $8\frac{1}{2}$ in. The printed area

of the page is $9\frac{1}{2}$ in. by $7\frac{1}{4}$ in. What is the area of the

margin?

24.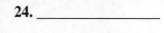

25. A room is 20 ft by 38 ft. The ceiling is 10 ft above the floor. There are six windows in the room, each 3 ft by 6 ft.

There are two doors in the room, each $2\frac{1}{2}$ ft by $6\frac{1}{2}$ ft.

a) What is the total area of the walls? Of the ceiling?

25. a)_____

b) A gallon of paint will cover $86\frac{5}{8}$ ft². How many

gallons of paint are needed for the walls? For the ceiling? (You must buy whole gallons. Assume a different color is used for the walls than for the ceiling.)

b)_____

c) Paint costs $28.50 a gallon. How much will it cost to paint the room?

c)_____

Find the area of the shaded region.

26.

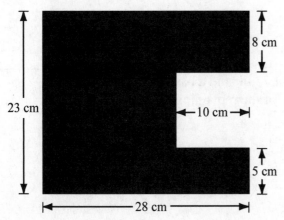

26. _____

27.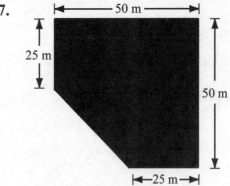

27. _____

28.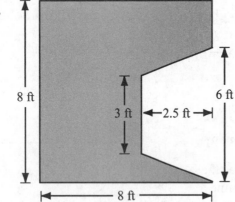

28. _____

29.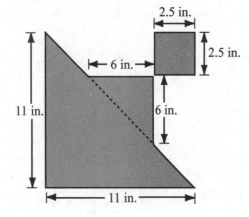

29. _____

30. A rectangular piece of cloth is 36 in. by 120 in. A triangular piece of cloth with a height of 72 in. and a base of 24 in. is cut from it. What is the area of the remaining cloth?

30. _____

Chapter 9 GEOMETRY

9.3 Circles

Learning Objectives
a   Find the length of a radius of a circle given the length of a diameter, and find the length of a diameter given the length of a radius.
b   Find the circumference of a circle given the length of a diameter or a radius.
c   Find the area of a circle given the length of a radius.
d   Solve applied problems involving circles.

Key Terms
Use the vocabulary terms listed below to complete each statement in Exercises 1–3. Terms may be used more than once.

radius   **diameter**   **circumference**

1.   A _____ is a segment that passes through the center of a circle and has endpoints on the circle.

2.   The _____ of a circle is the distance around it.

3.   The length of the _____ of a circle is half the length of the

_____ of the circle.

Objective a   Find the length of a radius of a circle given the length of a diameter, and find the length of a diameter given the length of a radius.

Find the length of the diameter.

4.   4. _____

5.   5. _____

6.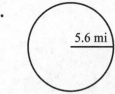

5.6 mi

Find the length of the radius.

7.

30.4 yd

7. _____

8.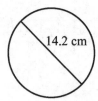

14.2 cm

8. _____

9.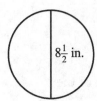

$8\frac{1}{2}$ in.

9. _____

Objective b   Find the circumference of a circle given the length of a diameter or a radius.

10.–15. Find the circumference of each circle in Exercises 4–9. Use $\dfrac{22}{7}$ for π in Exercises 4–6 and use 3.14 for π in Exercises 7–9.

10. _____

11. _____

178

12. _____

13. _____

14. _____

15. _____

Objective c   Find the area of a circle given the length of a radius.

16.–21. Find the area of each circle in Exercises 4–9. Use $\dfrac{22}{7}$ for π in Exercises 4–6 and use 3.14 for π in Exercises 7–9.

16. _____

17. _____

18. _____

179

19. _____

20. _____

21. _____

Objective d   Solve applied problems involving circles.

Solve. Use 3.14 for π.

22. The top of a can of fruit has a radius of 4 cm. What is the   22. _____
    diameter? the circumference? the area?

23. Which is larger and by how much: two round 8 in.   23. _____
    diameter cakes or one 9 in. by 13 in. rectangular cake?
    (Assume equal thickness.)

24. A round flower bed has a diameter of 6 ft. What is the   24. _____
    circumference? the area?

180

25. The circumference of a tree is 4 ft. What is the tree's diameter?

25. _____

Find the perimeter. Use 3.14 for π.

26.   12 in.

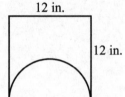

12 in.

26. _____

27.

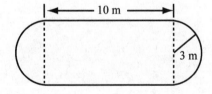

27. _____

28. Find the area of the figure in Exercise 26. Use 3.14 for π.

28. _____

29. Find the area of the figure in Exercise 27. Use 3.14 for π.

29. _____

30. Find the area of the shaded region. Use 3.14 for π.

30. _____

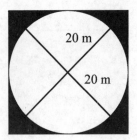

Chapter 9 GEOMETRY

9.4 Volume

Learning Objectives
a	Find the volume of a rectangular solid using the formula $V = l \cdot w \cdot h$.
b	Given the radius and the height, find the volume of a circular cylinder.
c	Given the radius, find the volume of a sphere.
d	Given the radius and the height, find the volume of a circular cone.
e	Solve applied problems involving volume of rectangular solids, circular cylinders, spheres, and cones.

Key Terms
Use the vocabulary terms listed below to complete each statement in Exercises 1–4.

rectangular solid   **circular cylinder**   **circular cone**   **sphere**

1. The volume of a _____ is given by $V = \frac{4}{3} \cdot \pi \cdot r^3$, where r is the radius.

2. The volume of a _____ is given by $V = l \cdot w \cdot h$, where l is the length, w is the width, and h is the height.

3. The volume of a _____ is given by $V = \pi \cdot r^2 \cdot h$, where r is the radius of the base and h is the height.

4. The volume of a _____ is given by $V = \frac{1}{3} \cdot \pi \cdot r^2 \cdot h$, where r is the radius of the base and h is the height.

Objective a   Find the volume of a rectangular solid using the formula $V = l \times w \times h$.

Find the volume of the rectangular solid.

5.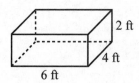

5. _____

6.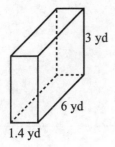

3 yd

6 yd

1.4 yd

6. _____

7.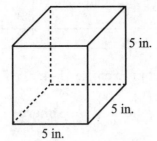

5 in.

5 in.

5 in.

7. _____

8.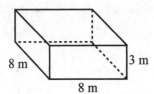

8 m

3 m

8 m

8. _____

9.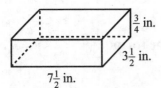

$\frac{3}{4}$ in.

$3\frac{1}{2}$ in.

$7\frac{1}{2}$ in.

9. _____

184

Objective b   Given the radius and the height, find the volume of a circular cylinder.

Find the volume of the circular cylinder. Use 3.14 for π in Exercises 10 and 11. Use $\frac{22}{7}$ for π in Exercises 12 and 13.

10.

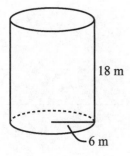

18 m

6 m

10. _____

11.

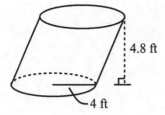

4.8 ft

4 ft

11. _____

12.

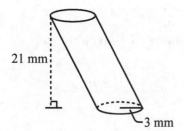

21 mm

3 mm

12. _____

13.

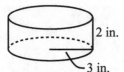

2 in.

3 in.

13. _____

Objective c   Given the radius, find the volume of a sphere.

Find the volume of the sphere. Use 3.14 for π *in Exercises 14 and 15. Use* $\dfrac{22}{7}$ *for* π *in Exercises 16 and 17.*

14.

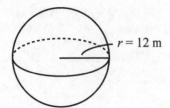

$r = 12$ m

14. _____

15.

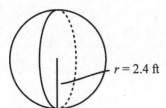

$r = 2.4$ ft

15. _____

16.

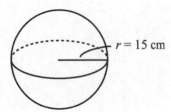

$r = 15$ cm

16. _____

17.

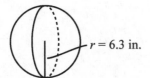

$r = 6.3$ in.

17. _____

Objective d   Given the radius and the height, find the volume of a circular cone.

Find the volume of the circular cone. Use 3.14 for π *in Exercises 18 and 19. Use* $\dfrac{22}{7}$ *for* π *in Exercises 20 and 21.*

18.

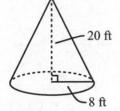

20 ft

8 ft

18. _____

186

19.

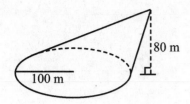

19. _____

20.

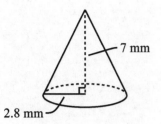

20. _____

21.

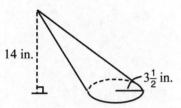

21. _____

Objective e   Solve applied problems involving volume of rectangular solids, circular cylinders, spheres, and cones.

Solve.

22. A pole has a diameter of 6 in. and a height of 5 ft (60 in.). Find its volume. Use 3.14 for π.

22. _____

23. A barrel has a height of $2\frac{1}{2}$ ft and a diameter of $1\frac{1}{2}$ ft. Find its volume. Use 3.14 for π.

23. _____

24. A ball has a diameter of 10 in. Find its volume. Use 3.14 for π.

24. _____

25. Pluto has a radius of about 742 mi. Find its volume. Use 3.14 for π.

25. _____

26. A party hat shaped like a cone has a height of 8 in. and the radius of its base is 3 in. Find its volume. Use 3.14 for π.

26. _____

27. A capsule of cold medicine has a length of 1 cm and a diameter of 0.2 cm. Find its volume. Use 3.14 for π. (Hint: First find the length of the cylindrical section.) Round to the nearest thousandth.

27. _____

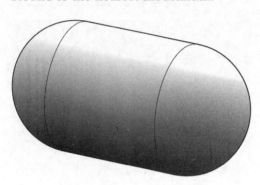

28. Kelly purchased a Rubik's Cube® that was 3 in. on each side. Find its volume.

28. _____

29. A box is just big enough to hold a spherical truffle. If the radius of the truffle is 1.4 cm, how much air surrounds the truffle? Use 3.14 for π and round to the nearest hundredth.

29. _____

30. The circumference of a ball at its widest point is 100π. Find its radius.

30. _____

Chapter 9 GEOMETRY

9.5 Angles and Triangles

Learning Objectives
a   Name an angle in five different ways, and given an angle, measure it with a protractor.
b   Classify an angle as right, straight, acute, or obtuse.
c   Identify complementary and supplementary angles and find the measure of a complement or a supplement of a given angle.
d   Classify a triangle as equilateral, isosceles, or scalene and as right, obtuse, or acute.
e   Given two of the angle measures of a triangle, find the third.

Key Terms
Use the vocabulary terms listed below to complete each statement in Exercises 1–7.

acute	**complementary**	**obtuse**	**right**
straight	**supplementary**	**ray**	

1.   Two angles are _____ if the sum of their measures is 180°.

2.   Two angles are _____ if the sum of their measures is 90°.

3.   A _____ has exactly one endpoint.

4.   A(n) _____ angle measures 180°.

5.   A(n) _____ angle measures 90°.

6.   A(n) _____ angle measures more than 0° but less than 90°.

7.   A(n) _____ angle measures more than 90° but less than 180°.

Objective a   Name an angle in five different ways, and given an angle, measure it with a protractor.

Name the angle in five different ways.

8.   8. _____

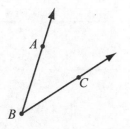

9.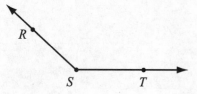

Use a protractor to measure the angle.

10.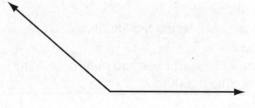

10. _____

11.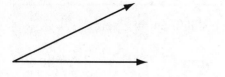

11. _____

12.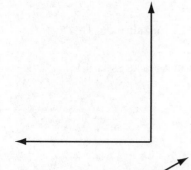

12. _____

13.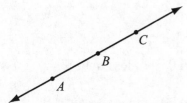

13. _____

Objective b   Classify an angle as right, straight, acute, or obtuse.

14.–17. Classify each of the angles in Exercises 10–13 as right, straight, acute, or obtuse.

14. _____

15. _____

16. _____

17. _____

Objective c   Identify complementary and supplementary angles and find the measure of a complement or a supplement of a given angle.

Find the measure of a complement of an angle with the given measure.

18. 17°   18. _____

19. 38°   19. _____

20. 79°   20. _____

Find the measure of a supplement of an angle with the given measure.

21. 8°   21. _____

22. 67°   22. _____

23. 123°   23. _____

Objective d   Classify a triangle as equilateral, isosceles, or scalene and as right, obtuse, or acute.

Classify each triangle as equilateral, isosceles, or scalene. Then classify it as right, obtuse, or acute.

24.   24. _____

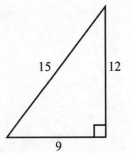

25.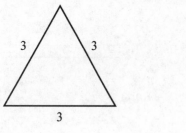

25. _____

26.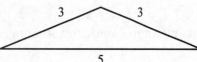

26. _____

27.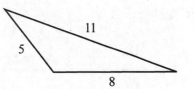

27. _____

Objective e   Given two of the angle measures of a triangle, find the third.

Find each missing angle measure.

28.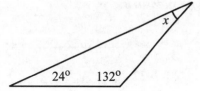

28. _____

29.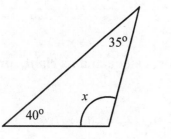

29. _____

30.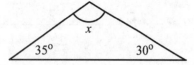

30. _____

31.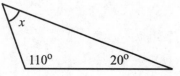

31. _____

192

Chapter 9 GEOMETRY

9.6 Square Roots and the Pythagorean Theorem

Learning Objectives

a   Simplify square roots of squares, such as $\sqrt{25}$.
b   Approximate square roots.
c   Given the lengths of any two sides of a right triangle, find the length of the third side.
d   Solve applied problems involving right triangles.

Key Terms

Use the vocabulary terms listed below to complete each statement in Exercises 1–3. Some of the terms will not be used.

hypotenuse   **leg**   **square**   **square root**   **Pythagorean**

1.   The longest side of a right triangle is the _____.

2.   For a right triangle, the equation $a^2 + b^2 = c^2$ is called the _____ equation.

3.   If $c^2 = a$, then c is a _____ of a.

Objective a   Simplify square roots of squares, such as $\sqrt{25}$.

Simplify.

4. $\sqrt{36}$ 4. _____

5. $\sqrt{144}$ 5. _____

6. $\sqrt{1600}$ 6. _____

7. $\sqrt{400}$ 7. _____

8. $\sqrt{441}$ 8. _____

9. $\sqrt{8100}$

9. _____

Objective b   Approximate square roots.

Approximate to three decimal places.

10. $\sqrt{13}$

10. _____

11. $\sqrt{10}$

11. _____

12. $\sqrt{31}$

12. _____

13. $\sqrt{58}$

13. _____

14. $\sqrt{136}$

14. _____

15. $\sqrt{513}$

15. _____

16. $\sqrt{222}$

16. _____

Objective c   Given the lengths of any two sides of a right triangle, find the length of the third side.

Find the length of the third side of each right triangle. Give an exact answer and, where appropriate, an approximation to three decimal places.

17.

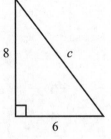

17. _____

194

18.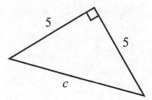

18. _____

19.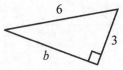

19. _____

20.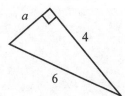

20. _____

For each right triangle, find the length of the side not given. Assume that c represents the length of the hypotenuse. Give an exact answer and, when appropriate, an approximation to three decimal places.

21. $a = 5$, $c = 13$

21. _____

22. $b = 12$, $c = 20$

22. _____

23. $a = 4$, $b = 7$

23. _____

24. $a = 10$, $b = 3$

24. _____

Objective d   Solve applied problems involving right triangles.

In Exercises 25–27, give an exact answer and an approximation to the nearest tenth.

25. How long must a wire be in order to reach from the top of   **25.** _____
a 12-ft pole to a point on the ground 4 ft from the base of
the pole?

26. How long is a rope reaching from the top of a 2 m pole to   **26.** _____
a point on the ground 1 m from the base of the pole?

27. How tall is this pole?   **27.** _____

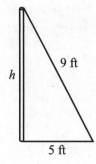

h

9 ft

5 ft

196

Chapter 10 REAL NUMBERS

10.1 The Real Numbers

Learning Objectives
a   State the integer that corresponds to a real world situation.
b   Graph rational numbers on a number line.
c   Convert from fraction notation to decimal notation for a rational number.
d   Determine which of two real numbers is greater and indicate which, using < or >.
e   Find the absolute value of a real number.

Key Terms
Use the vocabulary terms listed below to complete each statement in Exercises 1–8.

absolute value	**graph**	**integers**	**natural numbers**
opposites	**rational numbers**	**set**	**whole numbers**

1. A(n) _____ is a collection of objects.

2. We call –1 and 1 _____ of each other.

3. To _____ a number means to find and mark its point on a number line.

4. The _____ of a number is its distance from zero on a number line.

5. The set of _____ = {1, 2, 3,...}.

6. The set of _____ = {0, 1, 2, 3,...}.

7. The set of _____ = {..., –4, –3, –2, –1, 0, 1, 2, 3, 4,...}.

8. The set of _____ = the set of numbers $\frac{a}{b}$, where a and b are integers

    and $b \neq 0$.

Objective a   State the integer that corresponds to a real world situation.

State the integers that correspond to the situation.

9. On March 12, the temperature was 3° below zero. On   9. _____
    March 21, it was 55° above zero.

10. Will withdrew $480 from his savings account to buy textbooks. The next day, he deposited his paycheck of $325.

10. _____

Objective b   Graph rational numbers on a number line.

Graph the number on the number line.

11. $-\dfrac{7}{4}$

11. _____

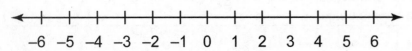

12. 2.67

12. _____

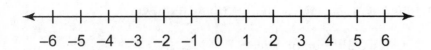

Objective c   Convert from fraction notation to decimal notation for a rational number.

Convert to decimal notation.

13. $-\dfrac{3}{8}$

13. _____

14. $\dfrac{7}{3}$

14. _____

15. $\dfrac{13}{6}$

15. _____

16. $-\dfrac{8}{9}$

16. _____

198

17. $-\dfrac{5}{4}$

17. _____

18. $\dfrac{3}{20}$

18. _____

Objective d   Determine which of two real numbers is greater and indicate which, using < or >.

Use either < or > for □ to write a true sentence.

19.   4 □ 0

19. _____

20.   −3 □ 1

20. _____

21.   14 □ −14

21. _____

22.   0 □ −6

22. _____

23.   −8 □ −9

23. _____

24.   −4.6 □ −1.2

24. _____

25. $\dfrac{3}{4}$ □ $-\dfrac{5}{4}$

25. _____

26. $\dfrac{2}{5}$ □ $\dfrac{3}{8}$

26. _____

Objective e   Find the absolute value of a real number.

Find the absolute value.

27. $|-16|$

27. _____

28. $|12|$

28. _____

29. $\left|-\dfrac{1}{4}\right|$

29. _____

30. $|-12.3|$

30. _____

31. $\left|4\dfrac{7}{8}\right|$

31. _____

32. $\left|\dfrac{0}{5}\right|$

32. _____

Chapter 10 REAL NUMBERS

10.2 Addition of Real Numbers

Learning Objectives
a   Add real numbers without using a number line.
b   Find the opposite, or additive inverse, of a real number.

Key Terms
Use the vocabulary terms listed below to complete each statement in Exercises 1–4.

     additive inverses   **negative**

     positive   **zero**

1.   The sum of two positive integers is _____ .

2.   The sum of two negative integers is _____ .

3.   If $a+b=a$, then b must be _____ .

4.   Two numbers whose sum is 0 are called _____ .

Objective a   Add real numbers without using a number line.

Add. Do not use a number line except as a check.

5. $6+(-11)$ 5. _____

6. $-8+3$ 6. _____

7. $-7+7$ 7. _____

8. $27+(-27)$ 8. _____

9. $-13+0$

9. _____

10. $-3+12$

10. _____

11. $50+(-25)$

11. _____

12. $-2+(-17)$

12. _____

13. $-1.3+(-3.8)$

13. _____

14. $\dfrac{2}{9}+\left(-\dfrac{7}{9}\right)$

14. _____

15. $-\dfrac{7}{6}+\dfrac{1}{3}$

15. _____

16. $-\dfrac{3}{20}+\dfrac{7}{15}$

16. _____

17. $-\dfrac{5}{24}+\left(-\dfrac{7}{18}\right)$

17. _____

202

18. $-\dfrac{3}{20}+\left(-\dfrac{7}{24}\right)$ 18. _____

19. $-32+\left(-\dfrac{5}{11}\right)+108+\left(-\dfrac{6}{11}\right)$ 19. _____

20. $36+(-45)+657+(-506)+(-89)$ 20. _____

Objective b   Find the opposite, or additive inverse, of a real number.

Find the opposite, or additive inverse.

21. -48 21. _____

22. 16.3 22. _____

Evaluate $-x$ when:

23. $x=\dfrac{11}{16}$ 23. _____

24. $x=-54$ 24. _____

Evaluate –(–x) when:

25. $x = -19$

25. _____

26. $x = 8.3$

26. _____

Find the opposite. (Change the sign.)

27. 28

27. _____

28. $-\dfrac{7}{3}$

28. _____

204

Chapter 10 REAL NUMBERS

10.3 Subtraction of Real Numbers

> **Learning Objectives**
> a   Subtract real numbers and simplify combinations of additions and subtractions.
> b   Solve applied problems involving subtraction of real numbers.

Key Terms
Use the vocabulary terms listed below to complete each statement in Exercises 1–2.

> **difference**   **opposite**

1.   The _____ $a - b$ is the number c for which $a = b + c$.

2.   To subtract, add the _____ of the number being subtracted.

Objective a   Subtract real numbers and simplify combinations of additions and subtractions.

Subtract.

3. $4 - 14$ 3. _____

4. $-10 - (-10)$ 4. _____

5. $-2 - (-17)$ 5. _____

6. $13 - 20$ 6. _____

7. $3 - (-3)$ 7. _____

8. $18 - (-40)$ 8. _____

9. $0 - (-7)$ 9. _____

10. $-16 - 0$ 10. _____

11. $-\dfrac{1}{6}-\dfrac{11}{12}$

12. $-\dfrac{1}{9}-\left(-\dfrac{5}{6}\right)$

13. $2.6-(-5.2)$

14. $8-4.63$

15. $3-(-4.75)$

16. $0-(-200)$

Simplify.

17. $32-(-25)-36+17-9$

18. $64-98+(-17)-(-38)-95$

19. $-1.7-(-15.7)+(-12.3)-(-16)$

20. $-\dfrac{2}{9}+\dfrac{5}{6}-\left(-\dfrac{7}{12}\right)$

Objective b   Solve applied problems involving subtraction of real numbers.

Solve.

21. From an elevation of 38 ft below sea level, Devin climbed to an elevation of 92 ft above sea level. How much higher was Devin at the end of his climb than at the beginning?

22. On January 10, the temperature fell from 12°F to –8°F. By how much did the temperature drop?

11. _____

12. _____

13. _____

14. _____

15. _____

16. _____

17. _____

18. _____

19. _____

20. _____

21. _____

22. _____

206

Chapter 10 REAL NUMBERS

10.4 Multiplication of Real Numbers

Learning Objectives
a   Multiply real numbers.

Key Terms
Use the vocabulary terms listed below to complete each statement in Exercises 1-4. Each term will be used more than once.

negative   **positive**

1.   The product of a positive number and a negative number is _____ .

2.   The product of two negative numbers is _____ .

3.   The product of an even number of negative numbers is _____ .

4.   The product of an odd number of negative numbers is _____ .

Objective a   Multiply real numbers.

Multiply.

5. $-5 \cdot 4$ 5. _____

6. $6 \cdot (-5)$ 6. _____

7. $-8 \cdot (-7)$ 7. _____

8. $-11 \cdot 12$ 8. _____

9. $-6.7 \cdot (-32)$ 9. _____

10. $-84 \cdot (-4.5)$

11. $-\dfrac{5}{9} \cdot \left(\dfrac{3}{10}\right)$

12. $-\dfrac{4}{5} \cdot \left(\dfrac{7}{8}\right)$

13. $-2 \cdot (-5) \cdot (-8) \cdot (-1)$

14. $0.02 \cdot (-4) \cdot 8 \cdot (-5)$

15. $5 \cdot (-4) \cdot 3 \cdot (-2) \cdot 1 \cdot 0$

16. $(-2)(-3)(-4)(-5)(-6)$

17. Evaluate $(-5x)^2$ and $-5x^2$ when $x = -4$.

18. Evaluate $3x^2$ when $x = 6$ and when $x = -6$.

10. _____

11. _____

12. _____

13. _____

14. _____

15. _____

16. _____

17. _____

18. _____

Chapter 10 REAL NUMBERS

10.5 Division of Real Numbers and Order of Operations

Learning Objectives
a   Divide integers.
b   Find the reciprocal of a real number.
c   Divide real numbers.
d   Solve applied problems involving multiplication and division of real numbers.
e   Simplify expressions using rules for order of operations.

Key Terms

Use the vocabulary terms listed below to complete each statement in Exercises 1-4.

negative   not defined   positive   reciprocals

1.   Division by 0 is _____ .

2.   The quotient of two numbers with the same sign is _____ .

3.   The quotient of two numbers with different signs is _____ .

4.   Two numbers whose product is 1 are called _____ .

Exercises 5–8 list the rules for order of operations. Use the vocabulary terms listed below to complete each statement.

additions   divisions   exponential   grouping

5.   Do all calculations within _____ symbols before operations outside.

6.   Evaluate all _____ expressions.

7.   Do all multiplications and _____ in order from left to right.

8.   Do all _____ and subtractions in order from left to right.

Objective a   Divide integers.

Divide, if possible. Check each answer.

9. $36 \div (-4)$

9.   _____

10. $\dfrac{26}{-2}$

10. _____

11. $\dfrac{-75}{-3}$

11. _____

12. $-66 \div (-6)$

12. _____

13. $\dfrac{66}{0}$

13. _____

14. $\dfrac{0}{-3}$

14. _____

Objective b   Find the reciprocal of a real number.

Find the reciprocal.

15. $\dfrac{4}{9}$

15. _____

16. $-\dfrac{6}{17}$

16. _____

17. -12

17. _____

18. 4.6

18. _____

19. $\dfrac{-1}{3t}$

19. _____

20. $\dfrac{-2x}{3y}$

20. _____

Objective c   Divide real numbers

Rewrite the division as a multiplication.

21. $3 \div 8$

21. _____

22. $\dfrac{5}{-9}$

22. _____

210

23. $-\dfrac{6.8}{1.3}$

 _____

24. $\dfrac{c}{\dfrac{1}{d}}$

 _____

25. $\dfrac{2x-1}{4}$

 _____

26. $\dfrac{3a+a^2}{a+1}$

 _____

Divide.

27. $\dfrac{5}{6}\div\left(-\dfrac{7}{5}\right)$

 _____

28. $-\dfrac{3}{8}\div\left(-\dfrac{6}{7}\right)$

 _____

29. $-\dfrac{2}{5}\div\left(-\dfrac{4}{9}\right)$

 _____

30. $-\dfrac{2}{3}\div\left(-\dfrac{3}{2}\right)$

 _____

31. $-10.4\div2.6$

 _____

32. $\dfrac{-2}{11-11}$

Objective d   Solve applied problems involving multiplication and division of real numbers.

Solve.

33. Vivian lost 1.5 lb each week for a period of 8 weeks. Express her total weight change as an integer.

33. _____

34. The temperature of a chemical compound was –10°C at 4:30 p.m. During a reaction, it dropped 2°C per minute until 4:38 p.m.. What was the temperature at 4:38 p.m.?

34. _____

35. The price of a stock began the day at $42.75 per share and dropped $0.38 per hour for 8 hours. What was the price of the stock after 8 hr?

35. _____

36. After diving 55 m below sea level, a diver rises at a rate of 6 m per minute for 8 min. Where is the diver in relation to the surface?

36. _____

A percent of increase is generally positive and a percent of decrease is generally negative. In Exercises 37 and 38, find the missing numbers. Round to the nearest tenth of a percent.
(Source: Veronis Shuler Stevenson, New York, New York)

37.

37. _____

Hours Per Person Spent Playing Video Games			
2000	2008	Change	Percent of Increase or Decrease
59	98	39	

38.

Hours Per Person Spent Reading Magazines			
2000	2008	Change	Percent of Increase or Decrease
135	110	−25	

38. _____

Objective e   Simplify expressions using rules for order of operations.

Simplify.

39. $4 - 3 \cdot 5 - 6$

39. _____

40. $\left[(-28) \div (-7) \right] \div \left(-\dfrac{1}{2} \right)$

40. _____

41. $13 - 3^2 + 3^3$

41. _____

42. $2^4 + 13 \cdot 16 - (12 + 32 \cdot 3)$

42. _____

43. $6 \cdot 7 - 2 \cdot 4 + 10$

43. _____

44. $10^2 / 20$

44. _____

45. $35 - 4(-5) + 8$

45. _____

46. $15 \div (-3) + 20 \div 2$

46. _____

47. $-7^2 - 8$

47. _____

48. $26 - 30^2$

48. _____

49. $3 \cdot 5^2 - 65$

49. _____

50. $\dfrac{4 - 3^2}{2^3 + 2^2}$

50. _____

51. $\dfrac{2(8-9) - 3 \cdot 4}{5 \cdot 6 - 3(8-1)}$

51. _____

Chapter 11 ALGEBRA: SOLVING EQUATIONS AND PROBLEMS

11.1 Introduction to Algebra

Learning Objectives
a   Evaluate algebraic expressions by substitution.
b   Use the distributive laws to multiply expressions like 8 and $x - y$.
c   Use the distributive laws to factor expressions like $4x - 12 + 24y$.
d   Collect like terms.

Key Terms
Use the vocabulary terms listed below to complete each statement in Exercises 1–8.

algebraic expression	**constant**	**equivalent**	**evaluating**
substituting	**value**	**factor**	**variable**

1.   A combination of letters, numbers, and operation signs, such as $16x - 18y$ is called a(n)

    _____ .

2.   A letter that can represent various numbers is a(n) _____ .

3.   A letter that can stand for just one number is a(n) _____ .

4.   When we replace a variable with a number, we are _____ for the

    variable.

5.   When we replace all variables in an expression with numbers and carry out the

    operations, we are _____ the expression.

6.   The results of evaluating an algebraic expression is called the _____ of

    the expression.

7.   Two expressions that have the same value for all allowable replacements are called

    _____ expressions.

8. To _____ an expression, we find an equivalent expression that is a product.

Objective a Evaluate algebraic expressions by substitution.

Substitute to find values of the expressions in each of the following applied problems.

9. The area A of a triangle with base b and height h is given by $A = \frac{1}{2}bh$. Find the area when $b = 32$ cm (centimeters) and $h = 15$ m.

9.

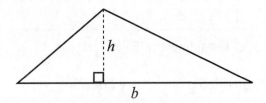

10. A driver who drives at a constant speed of r mph for t hr will travel a distance d mi given by $d = rt$ mi. How far will a driver travel at the speed of 70 mph for 3 hr?

10.

11. The simple interest on a principal of P dollars at interest rate r for time t, in years, is given by $I = Prt$. Find the simple interest on a principal of $1500 at 6% for 3 yr. (Hint: 6% = 0.06)

11.

12. A rectangular piece of paper is 5 in. wide and 9 in. long. Find its area.

12. _____

216

Evaluate.

13. $10z$, when $z = 3$

13. _____

14. $\dfrac{x}{y}$, when $x = 42$ and $y = 6$

14. _____

15. $\dfrac{5p}{q}$, when $p = 8$ and $q = 10$

15. _____

16. $\dfrac{a-b}{3}$, when $a = 25$ and $b = 13$

16. _____

Objective b   Use the distributive laws to multiply expressions like 8 and $x - y$.

Multiply.

17. $3(2x+8)$

17. _____

18. $5(x+7+2n)$

18. _____

19. $11(p-3)$

19. _____

20. $\dfrac{3}{4}(y-12)$

20. _____

21. $-4(-6x-y+9)$

21. _____

22. $-1.4(-2.3a-1.6b-3.4)$

22. _____

Objective c   Use the distributive laws to factor expressions like $4x - 12 + 24y$.

Factor. Check by multiplying.

23. $3x + 18$

23. _____

24. $12y + 30z$

24. _____

25. $9m - 45$

25. _____

26. $-5t + 20$

26. _____

27. $-21c + 35d + 7$

27. _____

28. $\dfrac{3}{4}x - \dfrac{7}{4}y + \dfrac{1}{4}$

28. _____

Objective d   Collect like terms.

Collect like terms.

29. $14x - x$

29. _____

30. $8t - 11t$

30. _____

31. $4m^2 - 10m - 5m^2$

31. _____

32. $16 - 40c - 20 - 5c + 6 + 8c$

32. _____

33. $-4 + 8a - 5b + 36a - 17b + 12$

33. _____

34. $5y + z + \dfrac{1}{3}y - \dfrac{3}{5}z - 11$

34. _____

218

Chapter 11 ALGEBRA: SOLVING EQUATIONS AND PROBLEMS

11.2 Solving Equations: The Addition Principle

Learning Objectives
a   Solve equations using the addition principle.

Key Terms
Use the vocabulary terms listed below to complete each statement in Exercises 1–2.

addition principle   **equivalent equations**

1.   Equations with the same solutions are called _____ .

2.   The _____ states that for any real numbers a, b, and c, $a = b$ is

       equivalent to $a + c = b + c$.

Objective a   Solve equations using the addition principle.

Solve using the addition principle. Don't forget to check!

3. $x + 8 = 20$ 3. _____

       Check: $x + 8 = 20$

4. $t + 12 = 42$ 4. _____

       Check: $t + 12 = 42$

5. $z + 3 = -8$

6. $y - 5 = 21$

7. $a - 3 = -5$

8. $-6 + x = 17$

9. $-8 + y = -15$

10. $a + \dfrac{2}{3} = 12$

11. $y - \dfrac{3}{5} = \dfrac{7}{8}$

12. $10.8 = x + 4.9$

13. $-2.9 = -1.4 + y$

13. _____

14. $6\dfrac{2}{3} = 4\dfrac{1}{2} + x$

14. _____

Chapter 11 ALGEBRA: SOLVING EQUATIONS AND PROBLEMS

11.3 Solving Equations: The Multiplication Principle

Learning Objective
a   Solve equations using the multiplication principle.

Key Terms
Use the vocabulary terms listed below to complete each statement in Exercises 1–4.

coefficient   **identity**   **inverse**   **principle**

1. The multiplication _____ states that for any real numbers a, b, and c, $c \neq 0$, $a = b$ is equivalent to $a \cdot c = b \cdot c$.

2. The multiplicative _____ of 3 is $\frac{1}{3}$.

3. The multiplicative _____ is 1 since $1 \cdot x = x$.

4. The _____ in $8x$ is 8.

Objective a   Solve equations using the multiplication principle.

Solve using the multiplication principle. Don't forget to check!

5. $4x = 28$ 5. _____

6. $3x = 36$ 6. _____

7. $-y = 20$ 7. _____

8. $-t = -5$ 8. _____

9. $11x = -66$

9. _____

10. $-9x = -108$

10. _____

11. $-32x = -96$

11. _____

12. $\dfrac{m}{8} = -7$

12. _____

13. $\dfrac{2}{5}y = 18$

13. _____

14. $\dfrac{-a}{7} = 8$

14. _____

15. $-\dfrac{3}{4}r = \dfrac{5}{8}$

15. _____

16. $-\dfrac{4}{5}a = -\dfrac{16}{15}$

16. _____

17. $5.9x = 17.7$

17. _____

18. $-3.6y = -21.6$

18. _____

19. $-\dfrac{3}{8}m = -22.14$

19. _____

20. $\dfrac{-x}{3} = -45$

20. _____

21. $\dfrac{t}{-6} = 8$

21. _____

Chapter 11 ALGEBRA: SOLVING EQUATIONS AND PROBLEMS

11.4 Using the Principles Together

Learning Objectives
a Solve equations using both the addition and the multiplication principles.
b Solve equations in which like terms may need to be collected.
c Solve equations by first removing parentheses and collecting like terms.

Key Terms
Use the vocabulary terms listed below to complete each statement in Exercises 1–2.

clear fractions   **distributive laws**

1. We remove parentheses in an equation by multiplying using the _____.

2. We multiply every term on both sides of an equation by the least common multiple of all

    denominators in order to _____ .

Objective a   Solve equations using both the addition and the multiplication principles.

Solve. Don't forget to check!

3. $3x + 5 = 29$ 3. _____

4. $6x - 5 = 37$ 4. _____

5. $4x + 5 = -39$ 5. _____

6. $-23 = 7 + 5y$ 6. _____

7. $-6x + 11 = 29$

7. _____

8. $-7x - 18 = -28\dfrac{1}{2}$

8. _____

Objective b   Solve equations in which like terms may need to be collected.

Solve.

9. $4x + 5x = 63$

9. _____

10. $6x + 5x = 132$

10. _____

11. $-3y - 4y = 28$

11. _____

12. $x + \dfrac{1}{2}x = 12$

12. _____

13. $7x - 1 = 15 - x$

13. _____

14. $8x + 7 = 3x + 12$

14. _____

226

15. $2 + 5z - 11 = 5z + 4 - z$

15. _____

16. $4y - 3 + 2y = 6y + 8 - y$

16. _____

Solve. Clear fractions or decimals first.

17. $\dfrac{3}{4}x + \dfrac{1}{2}x = 5x + \dfrac{1}{2} + \dfrac{1}{4}x$

17. _____

18. $\dfrac{1}{6} + 2y = 7y - \dfrac{5}{12}$

18. _____

19. $\dfrac{2}{5} + \dfrac{3}{5}x = \dfrac{8}{15} + \dfrac{1}{2}x + \dfrac{3}{2}$

19. _____

20. $3.6x + 14.7 = 0.3 - 1.2x$

20. _____

21. $4.07 - 0.61x = 0.82 - 5.16x$

21. _____

22. $\dfrac{2}{3}x - \dfrac{1}{4}x = \dfrac{3}{5}x + 1$

22. _____

Objective c   Solve equations by first removing parentheses and collecting like terms.

Solve.

23. $4(2t-5)=28$

23. _____

24. $3(2+5y)-10=11$

24. _____

25. $8-5(2x-3)=3$

25. _____

26. $3(m+2)=8(m+7)$

26. _____

27. $10(3t+2)=7(4t+6)$

27. _____

28. $13-(3x+4)=5(x+7)+x$

28. _____

29. $b+(b-4)=(b+3)-(b+1)$

29. _____

30. $2[5-3(4-x)]-7=3[2(5x-1)+8]-15$

30. _____

Chapter 11 ALGEBRA: SOLVING EQUATIONS AND PROBLEMS

11.5 Applications and Problem Solving

Learning Objective
a   Translate phrases to algebraic expressions.
b   Solve applied problems by translating to equations.

Key Terms
Use the vocabulary terms listed below to complete each statement in Exercises 1–5.

check	familiarize	solve	state	translate

1. To _____ yourself with a problem, read it carefully, choose a variable

    to represent the unknown, and make a drawing.

2. To _____ a problem into mathematical language, write an equation.

3. To _____ an equation, find all replacements that make the equation

    true.

4. Always _____ the answer in the original problem.

5. As a final problem-solving step, _____ the answer to the problem

    clearly.

Objective a   Translate phrases to algebraic expressions.

Translate each phrase to an algebraic expression. Use any letter for the variable unless directed otherwise.

6. Eight more than some number   6. _____

7. Twenty less than some number   7. _____

8. *b* divided by *x*

8. _____

9. *c* subtracted from *a*

9. _____

10. The product of two numbers

10. _____

11. Six multiplied by some number

11. _____

12. Ten more than seven times some number

12. _____

13. Five less than the product of two numbers

13. _____

14. Four times some number plus seven

14. _____

15. The sum of twice a number plus five times another number

15. _____

16. Raena drove a speed of 55 mph for *t* hours. How far did Raena drive?

16. _____

Objective b   Solve applied problems by translating to equations.

Solve.

17. A 60-in. board is cut into two pieces. One piece is four times the length of the other. Find the lengths of the pieces.

17. _____

230

18. Caedan purchased five copies of his favorite CD to give to his friends. If the total spent was $72.80, how much was one copy of the CD?

18. _____

19. In a recent year, New Hampshire had 117 women holding legislative office. This was 53 more than the number of women holding office in Maryland. How many women held legislative office in Maryland?
Source: U.S. Census Bureau

19. _____

20. A total of 2 in. of precipitation was recorded in East Lake City on May 11 and 12. The amount recorded on May 11 was three times the amount recorded on May 12. How much was recorded on May 12?

20. _____

21. The sum of three consecutive integers is 69. What are the numbers?

21. _____

22. The sum of three consecutive even integers is 198. What are the integers?

22. _____

23. A rectangle has a perimeter of 88 ft. The length is 2 ft more than twice the width. Find the dimensions of the rectangle.

23.  _____

24. Caitlyn paid $54.40 for a sweater during a 15%-off sale. What was the regular price?

24. 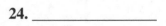 _____

25. Carlee paid $27.03, including 6% tax, for decorations for a party. What was the cost of the decorations before tax?

25.  _____

26. The second angle of a triangle is twice as large as the first angle. The third angle is 12° more than four times the first angle. How large are the angles?

26.  _____

27. The balance in Clayton's charge card account grew 3%, to $669.50, in one month. What was his balance at the beginning of the month?

27.  _____

232

28. The city garden has a rectangular patio. The length is 30 ft   **28.** _____
more than the width. The perimeter of the patio is 420 ft.
Find the length, the width, and the area of the patio.

29. A taxi cost $1.50 plus 60 ¢ per mile. How far can Carissa   **29.** _____
travel for $8.70?

30. Craig left an 18% tip for a meal. The total cost of the meal,   **30.** _____
including the tip, was $51.92. What was the cost of the
meal before the tip was added?

31. Cayla paid an average of $15 per book for three books.   **31.** _____
The price of one book was $1 more than another, and the
remaining book cost $12. What were the prices of the
other two books?

Chapter 1 WHOLE NUMBERS

Section 1.1

Key Terms
1. natural numbers

Objective a

3. 6 hundreds
7. 7

5. 6 thousands

Objective b

9. 9 thousands + 0 hundreds + 1 ten + 2 ones
11. 8 hundred thousands + 5 ten thousands + 1 thousand + 7 hundreds + 2 tens + 4 ones
13. 2 hundred thousands + 0 ten thousands + 8 thousands + 6 hundreds + 4 tens + 0 ones

Objective c

15. Three thousand, nine hundred five

17. Eight million, seven hundred fifty thousand, two hundred thirty-one

19. 2605

Section 1.2

Key Terms
1. perimeter

Objective a

3. 567
7. 11,550

5. 7766
9. 2670

Objective b

11. 12 yd

Section 1.3

Key Terms
1. difference

Objective a

3. 52   5. 1776
7. 262   9. 8072
11. 2864   13. 1545

Section 1.4

Key Terms
1. product

Objective a

3. 78   5. 3042
7. 22,672

Objective b

9. 144 sq mi

Section 1.5

Key Terms

1. dividend; divisor; quotient

Objective a

3. 1   5. 80 R3
7. 26 R5   9. 230 R2
11. 1627 R3

Section 1.6

Key Terms

1. equation

Objective a

3. 685,150   5. 685,000
7. 505,500

Objective b

9. $40 + 80 = 120$

11. $29,000 - 15,000 = 14,000$

13. $230 \times \$10 = \$2300$

Objective c

15. <

17. >

Section 1.7

Key Terms

1. variable

Objective a

3. 18

5. 14

Objective b

7. 68
11. 1872
15. 12
19. 797
23. 180

9. 9
13. 1299
17. 68
21. 23
25. 59

Section 1.8

Key Terms

1. translate

Objective a

3. 78 credits
7. 13 mph
11. 884 mi
15. $673
19. $90

5. 60 pea plants
9. 128 oz
13. 432 tiles
17. 25 gal
21. $16

Section 1.9

Key Terms

1. base

3. exponent

Objective a

5. 6^3

7. 11^4

Objective b

9. 9

11. 1296

Objective c

13. 20
17. 343
21. 18
25. 24

15. 8
19. 80
23. 60

Objective d

27. 0

29. 1460

Chapter 2 FRACTION NOTATION: MULTIPLICATION AND DIVISION

Section 2.1

Key Terms

1. multiple

3. divisible

Objective a

5. yes
9. 1, 2, 3, 4, 6, 8, 12, 16, 24, 32, 48, 96

7. 1, 2, 3, 4, 6, 12

Objective b

11. 15, 30, 45, 60, 75, 90, 105, 120, 135, 150

13. yes

Objective c

15. composite
19. composite

17. prime

238

Objective d

21. $2 \cdot 2 \cdot 2 \cdot 3 \cdot 5$

23. $3 \cdot 3 \cdot 5$

25. $2 \cdot 2 \cdot 5 \cdot 13 \cdot 17$

Section 2.2

Key Terms

1. even

3. divisible by 3

Objective a

5. 48, 130, 264, 26, 352, 900, 13,956

7. 48, 264, 352, 900, 13,956

9. 48, 264, 900, 13,956

11. 48, 264, 352

13. 130, 900

15. 78, 351, 54, 3054, 369, 15,000

17. 25, 980, 15,000

19. 112, 980

21. 351, 54, 369

Section 2.3

Key Terms

1. denominator

Objective a

3. Numerator: 4; denominator: 5

5. Numerator: 12; denominator: 7

7. $\dfrac{6}{15}$

9. a. $\dfrac{79}{100}$ b. $\dfrac{79}{21}$ c. $\dfrac{21}{100}$

Objective b

11. 0

13. Not defined

15. Not defined

Section 2.4

Key Terms

1. solve

3. check

Objective a

5. $\dfrac{9}{64}$

7. $\dfrac{4}{27}$

9. $\dfrac{8}{5}$

Objective b

11. $\dfrac{5}{16}$ lb of seed

13. 19 books

15. $\dfrac{9}{2}$ yards

17. $\dfrac{1}{6}$ cup

Section 2.5

Key Terms

1. canceling

3. simplest

Objective a

5. $\dfrac{25}{30}$

7. $\dfrac{55}{75}$

Objective b

9. 5

11. $\dfrac{3}{4}$

Objective c

13. $=$

15. $\neq$

17. $\neq$

Section 2.6

Objective a

1. $\dfrac{1}{5}$

3. $\dfrac{3}{4}$

5. 1

240

Objective b

7. $4550

9. $\dfrac{1}{6}$ cup

11. 135 mi

Section 2.7

Key Terms

1. reciprocal; $\dfrac{n}{0}$

Objective a

3. $\dfrac{4}{3}$

5. 5

7. $\dfrac{8}{5}$

Objective b

9. $\dfrac{11}{6}$

11. $\dfrac{7}{10}$

13. $\dfrac{5}{4}$

15. 1

Objective c

17. 3

19. 96

21. $\dfrac{3}{10}$

Objective d

23. 20 bowls

25. 24 L

27. $\dfrac{1}{16}$ m

Chapter 3 FRACTION NOTATION AND MIXED NUMERALS

Section 3.1

Key Terms

1. prime factorization

3. least common multiple; multiple

Objective a

5. 18
9. 132
13. 187
17. 144
21. 1500

7. 168
11. 105
15. 66
19. 120
23. 24 min after starting

Section 3.2

Key Terms

1. different denominators

3. like denominators

Objective a

5. $\dfrac{3}{4}$

7. $\dfrac{11}{10}$

9. $\dfrac{9}{8}$

Objective b

11. $\dfrac{17}{12}$ lb

13. $\dfrac{17}{24}$ lb   14. $\dfrac{35}{24}$ lb

15. $\dfrac{55}{32}$ in.

Section 3.3

Objective a

1. $\dfrac{1}{2}$

3. $\dfrac{13}{48}$

5. $\dfrac{2}{15}$

Objective b

7. >

9. <

11. <

Objective c

13. $\dfrac{1}{3}$

15. $\dfrac{49}{104}$

Objective d

17. $\dfrac{5}{12}$ mi

19. $\dfrac{5}{12}$ lb

21. $\dfrac{35}{48}$ cup

Section 3.4

Key Terms

1. mixed numerals; fraction notation

Objective a

3. $\dfrac{9}{2}$

5. $\dfrac{205}{8}$

7. $2\dfrac{3}{10}$

Objective b

9. $137\dfrac{4}{7}$

11. $114\dfrac{22}{65}$

13. $-42\dfrac{5}{12}$

Section 3.5

Objective a

1. $9\frac{1}{5}$

3. $5\frac{13}{14}$

5. $16\frac{7}{24}$

Objective b

7. $4\frac{3}{5}$

9. $5\frac{1}{6}$

11. $35\frac{13}{40}$

Objective c

13. $6\frac{7}{12}$ in.

15. $20\frac{5}{8}$ yd

17. $121\frac{3}{4}$ in.

Section 3.6

Objective a

1. $16\frac{5}{7}$

3. 38

5. $601\frac{1}{15}$

Objective b

7. $1\frac{1}{2}$

9. $2\frac{7}{9}$

Objective c

11. $61\frac{9}{32}$ ft^2

13. 32 mpg

15. 77°F

17. $26\frac{2}{3}$ tsp

19. 102 gal

244

Section 3.7

Key Terms

1. average

Objective a

3. $\dfrac{1}{210}$

5. $\dfrac{4}{3}$

7. 1

9. $32\dfrac{19}{28}$

11. $\dfrac{10}{27}$

13. $\dfrac{29}{96}$

15. $22\dfrac{11}{18}$

Objective b

17. 0

19. $\dfrac{1}{2}$

21. Answers may vary. 8

23. Answers may vary. 37

25. 8

27. $\dfrac{1}{2}$

29. 3

Chapter 4 DECIMAL NOTATION

Section 4.1

Key Terms

1. arithmetic numbers; whole numbers; fractions
3. decimal point

Objective a

5. Two and seven hundred eighty-nine thousandths
7. Thirty-nine and seventy-three thousandths
9. Five and eight hundred one ten-thousandths

Objective b

11. $\dfrac{29}{100}$

13. $\dfrac{205,004}{1000}$

15. 49.58

17. 5.81749

Objective c

19. 0.51

21. $\dfrac{7}{100}$

Objective d

23. 0.7

25. 55,000

27. 54,795

29. 54,795.06

Section 4.2

Key Terms

1. credit

3. place-value digits

Objective a

5. 404.57

7. 53.21

9. 493.0756

Objective b

11. 35.84

13. 4.5965

15. 7.71

17. 5.90461

Objective c

19. 114.92

21. 22,984.69

23. 407.281

Objective d

25. $1548.50

27. $222.02

Section 4.3

Key Terms

1. cents; dollar

3. $; dollars

Objective a

5. 6.44
9. 526.5
13. 873
17. 504.67

7. 0.558
11. 4.89234
15. 7.36244
19. 2.18286

Objective b

21. 3507¢
25. $0.45
29. 43,600,000,000

23. 239¢
27. $48.62

Section 4.4

Key Terms

1. quotient

3. average

Objective a

5. 1.62
9. 5.2
13. 0.058
17. 347.6

7. 2.48
11. 38
15. 84

Objective b

19. 0.93

21. 4762.5

Objective c

23. 1289.4
27. 257.17875

25. 60.205
29. 0.29

Section 4.5

Key Terms
1. repeating decimal

Objective a

3. 0.51
5. 0.85
7. 1.24
9. 0.$\overline{428571}$
11. 1.41$\overline{6}$

Objective b

13. 1.2; 1.24; 1.240
15. 0.4; 0.43; 0.429
17. 1.4; 1.42; 1.417
19. 0.3; 0.29; 0.293
21. 25.1 mpg
23. 22.4

Objective c

25. 2.058
27. 98.208
29. 33.6

Section 4.6

Key Terms

1. difference
3. quotient

Objective a

5. d
7. c
9. a
11. b
13. 2.3
15. 91
17. 3.6
19. 130
21. b
23. a
25. d
27. a
29. $700 ÷ $70 = 10 calculators

Section 4.7

Key Terms

1. variable

248

Objective a

3. 1207.8 mi
7. 104.8576 mi²
11. 4.8 in.
15. 28,730.46 yd²
19. $86.90
23. 0.367

5. 102.4 °F
9. 37.5 m
13. $777.39
17. 1776.25 ft²
21. $843.55
25. 269 min

Chapter 5 RATIO AND PROPORTION

Section 5.1

Objective a

1. $\dfrac{3}{4}$

3. $\dfrac{149}{150}$

5. $\dfrac{5\frac{3}{4}}{6\frac{5}{8}}$

7. $\dfrac{35.95}{29.99}$; $\dfrac{29.99}{35.95}$

Objective b

9. $\dfrac{4}{5}$

11. $\dfrac{2}{3}$

13. $\dfrac{24}{25}$

15. $\dfrac{2}{3}$

Section 5.2

Key Terms

1. rate

Objective a

3. 50 mph
7. 606 persons/sq mi

5. 17.4 mpg

Objective b

9. B: 7.46¢/oz; B: 7.47¢/oz; brand A has the lowest unit price
11. A: 37.38¢/oz; B: 35.57¢/oz; C: 35.31¢/oz; brand C has the lowest unit price

Section 5.3

Key Terms

1. proportional; ratio

3. cross products

Objective a

5. No

7. Yes

Objective b

9. $\dfrac{15}{8}$

11. 0.18

13. 5.125

15. $\dfrac{57}{7}$, or $8\dfrac{1}{7}$

Section 5.4

Objective a

1. $93\dfrac{1}{3}$ Cal

3. 12 adults

5. 48 trout

7. 108 pens

9. 495 min, or 8 hr 15 min

Section 5.5

Key Terms

1. shape

Objective a

3. $RQ = 9$, $RS = 22.8$

5. $QN = 18$

7. 75 ft

9. 39 m

Objective b

11. 3.75

13. $x = 10.5$; $y = 17.5$; $z = 7$

Chapter 6 PERCENT NOTATION

Section 6.1

Key Terms

1. percent

Objective a

3. $\dfrac{45}{100}$; $45 \times \dfrac{1}{100}$; 45×0.01

5. $\dfrac{11.1}{100}$; $11.1 \times \dfrac{1}{100}$; 11.1×0.01

Objective b

7. 0.23

9. 0.3176

11. 0.06

13. 8

15. 0.0045

17. 0.042

Objective c

19. 0.35

21. 93%

23. 250%

25. 0.1%

27. 48.76%

29. 75%

Section 6.2

Key Terms

1. fraction equivalent

Objective a

3. 17%

5. 30%

7. 62.5%, or $62\dfrac{1}{2}\%$

9. 93.75%, or $93\dfrac{3}{4}\%$

11. 76%

13. 6%

15. $33.\overline{3}\%$, or $33\dfrac{1}{3}\%$

17. $16.\overline{6}\%$, or $16\dfrac{2}{3}\%$

Objective b

19. $\dfrac{3}{8}$

21. $\dfrac{1}{250}$

251

23. $\dfrac{7}{2}$

25. $\dfrac{17}{25}$

27. $\dfrac{1}{4}$

29. $\dfrac{7}{200}$

Section 6.3

Key Terms

1. base

Objective a

3. $y = 15\% \times 60$

5. $25 = a \times 125$

7. $20 = 45\% \times z$

Objective b

9. 70

11. \$32

13. 90

15. 320%

17. $133.\bar{3}\%$, or $133\dfrac{1}{3}\%$

19. \$450

21. 7.5

23. \$55.64

Section 6.4

Key Terms

1. whole

Objective a

3. $\dfrac{28}{100} = \dfrac{a}{36}$

5. $\dfrac{N}{100} = \dfrac{2.2}{20}$

7. $\dfrac{20}{100} = \dfrac{19}{b}$

Objective b

9. 21

11. 45

13. 44%

15. 25%

17. \$36

19. $161.\bar{1}$, or $161\dfrac{1}{9}$

21. 250

252

Section 6.5

Key Terms

1. percent of decrease

Objective a

3. 125 students
5. $72
7. 37.5 mL plant food concentrate; 712.5 mL water
9. 52.8 items correct; 7.2 items incorrect
11. 62.5%

Objective b

13. 8%
15. 2.6%
17. $560; $392
19. $14.04; $58.50
21. a) 1330; b) 10.6%

Section 6.6

Key Terms

1. commission
3. sales tax

Objective a

5. $6.05; $114.05
7. $18,500

Objective b

9. $875
11. $8200

Objective c

13. $24; $96
15. $112; $78.40
17. 25%; $187.50

Section 6.7

Key Terms

1. compound interest
3. APR

Objective a

5. $256.50

7. $140.05; $14,340.05

Objective b

9. $811.20

11. $176,753.91

13. $539,130.22

Objective c

15. a) $100; b) $89.67 interest, $10.33 principal; c) $57.97 interest, $42.03 principal;
    d) at 13.9%, $31.70 less interest and more principal is paid than at 21.5%

Chapter 7 DATA, GRAPHS, AND STATISTICS

Section 7.1

Key Terms

1. mean

3. mode

Objective a

5. 10

7. 4.5

9. 23.5 mpg

11. 97

Objective b

13. 11

15. 3.05

17. 69.5 °F

Objective c

19. 5

21. $25, $30

23. 0

Objective d

25. Battery A: average ≈ 19.4; battery B: average ≈ 19.2; battery A is better

27. Flavor A: average ≈ 7.8; flavor B: average ≈ 7.6; flavor A tastes better

Section 7.2

Key Terms

1. pictograph

254

Objective a

3. 40.1 yr

5. Argentina, Denmark, Georgia

7. Average: $87.\overline{3}\%$; median: 93.7%; no modes exist

9. 5,529,320 people

11. 1890

13. Delaware

15. $51,312

Objective a

17. Nevada

19. Alaska and Hawaii

21. 6500 bottles

23. 2003

25. 2002 and 2003

Section 7.3

Key Terms

1. horizontal

Objective a

3. Airline pilot

5. 17 hours

7. About 6 yr

9. About 6 yr

Objective b

11. Bananas

13. 13.6 pounds

15. 339 sq ft

Objective c

17. 3 points

19. 2004 and 2005

Objective d

21. $1.52

23. 9%

25. $637.60

27. $63

Section 7.4

Key Terms

1. pie chart

Objective a

3. 9%

5. 30 pints

7. 66%

8. 16%

9. 24%

10. 32%

11. 1200 homes

12. 10,120 homes

13. 69%

14. 54%

Objective b

15.

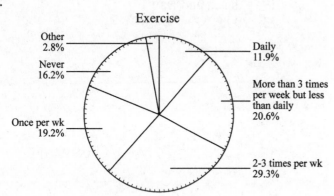

17.

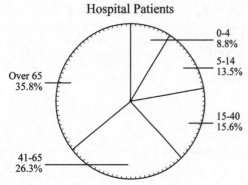

19.

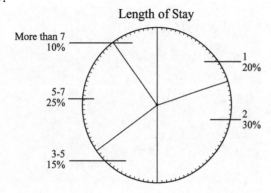

Chapter 8 MEASUREMENT

Section 8.1

Key Terms
1. 3

3. 12

Objective a

5. 15,840

7. 12

9. 90

11. 54

13. 4.5, or $4\frac{1}{2}$

Section 8.2

Key Terms

1. 100

3. 1000

Objective b

5. 40

6. 2500

7. 1400

9. 0.15

11. 270

13. 1.35

15. 4300

Section 8.3

Objective a

1. 12.7

3. 2.742

5. 78.74

7. About 70.9

9. 49.68

Section 8.4

Key Terms

1. 1000

3. 16

Objective a

5. 8000

7. 3.5

9. 40

11. 2.5

Objective b

13. 1.2

15. 200,000

17. 24

19. 6

21. 3600

Objective c

23. 3000

25. 0.05

27. 400 mcg

29. 0.018 mg

Section 8.5

Key Terms

1. 1

3. 16

Objective a

5. 4

7. 20

9. 128

11. 0.514

13. 7000

Objective b

15. 2.88 L

17. 0.42 g

19. 100 mL

Section 8.6

Key Terms

1. 212

Objective a

3. 4 min

5. 180 hr

7. 9.4 hr

9. 25,200 min

11. 3 wk

13. 1 wk

Objective b

15. 122 °F

17. 113 °F

19. 20 °C

21. 35 °C

Section 8.7

Key Terms

1. 144

Objective a

3. 27

5. 130,680

7. 5

9. 864

11. 5

Objective b

13. 457

15. 0.156

17. 2,000,000

19. 0.0000164976

Chapter 9 GEOMETRY

Section 9.1

Key Terms

1. polygon

3. square

Objective a

5. 84 m

7. 35 ft

9. 127 in.

11. $20\frac{1}{2}$ in.

13. 28 mi

15. $17\frac{1}{3}$ yd

Objective b

17. 54 ft

19. 34 ft

21. a) 40 posts; b) $128; c) 76 m; d) $209; e) $575

Section 9.2

Key Terms

1. $A = b \cdot h$

3. $A = \dfrac{1}{2} \cdot b \cdot h$

Objective a

5. 5.76 ft²

7. $68\dfrac{1}{16}$ mi²

9. 56.95 m²

11. 169 yd²

13. $\dfrac{9}{16}$ mi²

Objective b

15. 12.25 in²

17. 9.66 in²

19. 24 ft²

Objective c

21. 3510 m²

23. 404 ft²

25. a) $1019\dfrac{1}{2}$ ft², 760 ft²;

27. 2187.5 m²

   b) 12 gallons, 9 gallons; c) $598.50

29. 84.75 in²

Section 9.3

Key Terms

1. diameter

3. radius; diameter

Objective a

5. 28 ft

7. 15.2 yd

9. $4\dfrac{1}{4}$ in.

Objective b

11. 88 ft

13. 95.456 yd

15. 26.69 in.

260

Objective c

17. 616 ft²

19. 725.4656 yd²

21. 56.71625 in², or $56\frac{573}{800}$ in²

Objective d

23. 9 in. by 13 in. rectangle; 16.52 in²

25. About 1.27 ft

27. 38.84 m

29. 88.26 m²

Section 9.4

Key Terms

1. sphere

3. circular cylinder

Objective a

5. 48 ft³

7. 125 in³

9. $19\frac{11}{16}$ in³

Objective b

11. 241.152 ft³

13. $56\frac{4}{7}$ in³

Objective c

15. 57.87648 ft³

17. 1047.816 in³

Objective d

19. $837,333\frac{1}{3}$ m³

21. $179\frac{2}{3}$ in²

Objective e

23. $4\frac{133}{320}$ ft³, or 4.415625 ft³

25. 1,710,330,736 mi³

27. 0.036 cm³

29. 10.46 cm³

Section 9.5

Key Terms

1. supplementary
5. right

3. ray
7. obtuse

Objective a

9. Angle *RST*, angle *TSR*, ∠*RST*, ∠*TSR*, ∠*S*
13. 180°

11. 25°

Objective b

15. Acute

17. Straight

Objective c

19. 52°
23. 57°

21. 172°

Objective d

25. Equilateral; acute

27. Scalene; obtuse

Objective e

29. 105°

31. 50°

Section 9.6

Key Terms

1. hypotenuse

3. square root

Objective a

5. 12
9. 90

7. 20

Objective b

11. 3.162
15. 22.650

13. 7.616

262

Objective c

17. $c = 10$

19. $b = \sqrt{27};\ b \approx 5.196$

21. $b = 12$

23. $c = \sqrt{65};\ c \approx 8.062$

Objective d

25. $\sqrt{160}$ ft ≈ 12.6 ft

27. $\sqrt{56}$ ft ≈ 7.5 ft

Chapter 10 REAL NUMBERS

Section 10.1

Key Terms

1. set

3. graph

5. natural numbers

7. integers

Objective a

9. $-3;\ 55$

Objective b

11.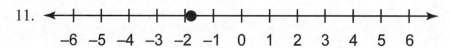

Objective c

13. -0.375

15. $2.1\overline{6}$

17. -1.25

Objective d

19. $>$

21. $>$

23. $>$

25. $>$

Objective e

27. 16

29. $\dfrac{1}{4}$

31. $4\dfrac{7}{8}$

Section 10.2

Key Terms

1. positive

3. zero

Objective a

5. −5

7. 0

9. −13

11. 25

13. −5.1

15. $-\dfrac{5}{6}$

17. $-\dfrac{43}{72}$

19. 75

Objective b

21. 48

23. $-\dfrac{11}{16}$

25. −19

27. −28

Section 10.3

Key Terms
1. difference

Objective a

3. −10

5. 15

7. 6

9. 7

11. $-\dfrac{13}{12}$

13. 7.8

15. 7.75

17. 29

19. 17.7

Objective b

21. 130 ft

Section 10.4

Key Terms

1. negative

3. positive

Objective a

5. -20

7. 56

9. 214.4

11. $-\dfrac{1}{6}$

13. 80

15. 0

17. $400; -80$

Section 10.5

Key Terms

1. not defined

3. negative

5. grouping

7. divisions

Objective a

9. -9

11. 25

13. Not defined

Objective b

15. $\dfrac{9}{4}$

17. $-\dfrac{1}{12}$

19. $-3t$

Objective c

21. $3 \cdot \left(\dfrac{1}{8} \right)$

23. $-6.8 \left(\dfrac{1}{1.3} \right),$ or $6.8 \cdot \left(-\dfrac{1}{1.3} \right)$

25. $(2x-1) \left(\dfrac{1}{4} \right)$

27. $-\dfrac{25}{42}$

29. $\dfrac{9}{10}$

31. -4

Objective d

33. -12 lb

35. \$39.71

37. 66.1%

Objective e

39. -17

41. 31

43. 44

45. 63

47. -57

49. 10

51. $-\dfrac{14}{9}$

Chapter 11 ALGEBRA: SOLVING EQUATIONS AND PROBLEMS

Section 11.1

Key Terms

1. algebraic expression

3. constant

5. evaluating

7. equivalent

Objective a

9. 240 cm^2

11. $270

13. 30

15. 4

Objective b

17. $6x+24$

19. $11p-33$

21. $24x+4y-36$

Objective c

23. $3(x+6)$

25. $9(m-5)$

27. $7(-3c+5d+1)$, or $-7(3c-5d-1)$

Objective d

29. $13x$

31. $-m^2-10m$

33. $44a-22b+8$

Section 11.2

Key Terms

1. equivalent equations

266

Objective a

3. 12
7. −2
11. $\dfrac{59}{40}$

5. −11
9. −7
13. −1.5

Section 11.3

Key Terms

1. principle

3. identity

Objective a

5. 7
9. −6
13. 45
17. 3
21. −48

7. −20
11. 3
15. $-\dfrac{5}{6}$
19. 59.04

Section 11.4

Key Terms

1. distributive laws

Objective a

3. 8
7. −3

5. −11

Objective b

9. 7
13. 2
17. $-\dfrac{1}{8}$
21. $-\dfrac{5}{7}$

11. −4
15. 13
19. $\dfrac{49}{3}$

Objective c

23. 6
27. 11

25. 2
29. 3

Section 11.5

Key Terms

1. familiarize
5. state

3. solve

Objective a

7. $t-20$
11. $6n$
15. $2c+5d$, or $5d+2c$

9. $a-c$
13. $st-5$

Objective b

17. 48 in.; 12 in.
21. 22, 23, 24
25. $25.50
29. 12 mi

19. 64 women
23. length: 30 ft; width: 14 ft
27. $650
31. $16, $17